02-WSM-2

NAVIGATING THE TMDL PROCESS: SEDIMENT TOXICITY

by:

Brian S. Anderson
John W. Hunt
Bryn M. Phillips
Ron S. Tjeerdema

University of California, Davis

Department of Environmental Toxicology

Marine Pollution Studies Laboratory

2007

WERF

The Water Environment Research Foundation, a not-for-profit organization, funds and manages water quality research for its subscribers through a diverse public-private partnership between municipal utilities, corporations, academia, industry, and the federal government. WERF subscribers include municipal and regional water and wastewater utilities, industrial corporations, environmental engineering firms, and others that share a commitment to cost-effective water quality solutions. WERF is dedicated to advancing science and technology addressing water quality issues as they impact water resources, the atmosphere, the lands, and quality of life.

For more information, contact:
Water Environment Research Foundation
635 Slaters Lane, Suite 300
Alexandria, VA 22314-1177
Tel: (703) 684-2470
Fax: (703) 299-0742
www.werf.org
werf@werf.org

This report was co-published by the following organizations. For non-subscriber sales information, contact:

IWA Publishing
Alliance House, 12 Caxton Street
London SW1H 0QS, United Kingdom
Tel: +44 (0) 20 7654 5500
Fax: +44 (0) 20 7654 5555
www.iwapublishing.com
publications@iwap.co.uk

Library of Congress Catalog Card Number: 2006928030
IWAP ISBN: 978-1-84339-766-3

University of California, Davis

The research on which this report is based was developed, in part, by the United States Environmental Protection Agency (EPA) through Cooperative Agreement No. # CR83155901 with the Water Environment Research Foundation (WERF). However, the views expressed in this document are solely those of University of California, Davis and neither EPA nor WERF endorses any products or commercial services mentioned in this publication. This report is a publication of WERF, not EPA. Funds awarded under the Cooperative Agreement cited above were not used for editorial services, reproduction, printing, or distribution.

ACKNOWLEDGMENTS

This research was funded by the Water Environment Research Foundation and the U.S. EPA. The authors would like to thank the following people for their assistance in various aspects of this project: Dave Crane, Abdou Mekebri, Jim McCall (Water Pollution Control Laboratory, Rancho Cordova, CA), Scott Cieniawski (U.S. EPA, Chicago, IL), David Moore and Brian Hester (Weston Solutions, Port, WA), Joy McGrath (HydroQual, Inc., Mahwah, NJ), and Howard Cumberland (CH2M Hill, Portland, OR).

Report Preparation

Principal Investigator:
Ron S. Tjeerdema, Ph.D., DBAT
University of California, Davis
Department of Environmental Toxicology

Project Team:
Brian S. Anderson, John W. Hunt, Bryn M. Phillips
University of California, Davis
Department of Environmental Toxicology

Project Subcommittee

Philip B. Dorn, Ph.D.
Shell Health Services

Kay Ho, Ph.D.
U.S. Environmental Protection Agency

Robert A. Hoke, Ph.D.
DuPont

Michael G. Kellogg
City and County of San Francisco

Drew McAvoy, Ph.D.
Procter & Gamble Company

Teresa J. Norberg-King, Ph.D.
U.S. Environmental Protection Agency

James Salerno
San Francisco Public Utilities Commission

Jeffery A Steevens, Ph.D.
U.S. Army Engineer Research and Development Center

Water Environment Research Foundation Staff

Director of Research: Daniel M. Woltering, Ph.D.
Program Manager: Jane M. Casteline, M.S.

ABSTRACT AND BENEFITS

Abstract:

A key step in the development of Total Maximum Daily Load (TMDL) allocations for waterbodies impaired due to sediment toxicity is the identification of chemicals responsible for toxicity. Sediment toxicity identification evaluation procedures (TIEs) are one of the primary tools used in this process. This project evaluated standardized and recently developed sediment TIE methods developed by the U.S. EPA and others, to determine their utility for identifying chemicals responsible for toxicity.

In this study, formulated sediments were spiked with five chemicals: copper, fluoranthene, tetrachlorobenzene, nonylphenol and ammonia. Experiments were conducted using sediments spiked with single chemicals and chemical mixtures. Toxicity tests used the estuarine amphipod *Eohaustorius estuarius* and the freshwater amphipod *Hyalella azteca*. TIEs were conducted using solid-phase (whole sediment) and sediment interstitial waters. The results indicate that the TIE methods are sufficient to characterize and identify toxicity due to single and multiple chemicals using spiked sediments. One key finding from the spiked sediment experiments is that methods to elute chemicals from extraction media used in solid-phase and interstitial water TIEs require further refinement to ensure complete and consistent elution of sorbed chemicals. An additional finding is that both solid-phase and interstitial water TIE procedures provide useful lines of evidence and that both approaches should be used in a weight-of-evidence approach in the sediment TIE process.

In the final phase of this project, the TIE procedures were evaluated using ambient samples collected from three marine sites and three freshwater sites. Marine sediments were collected from Switzer Creek in San Diego Bay, Upper Newport Bay in Newport Beach, and Consolidated Slip, in Long Beach Harbor, all of which are sites in southern California. Freshwater sediments were collected from two additional sites in California: San Diego Creek, in Newport Beach, and Alisal Creek, in Salinas. The final freshwater sediment was collected from a site in Indiana Harbor on the southwest shore of Lake Michigan in East Chicago, Indiana. Multiple solid-phase and interstitial water TIEs were conducted on each sediment and results of these were combined with chemical analyses and other lines-of-evidence to evaluate the methods. The results indicate that sediment TIE methods are sufficiently developed to characterize toxicity due to general classes of chemicals such as cationic metals, organic chemicals, and ammonia. Methods to improve extraction and elution of metal and organic chemicals in highly toxic sediments require further refinement in order to allow identification of specific chemicals responsible for toxicity as part of the Phase II TIE process. These refinements include determining the appropriate masses of solid-phase extraction media to allow complete removal of toxic chemicals from sediments and interstitial waters. In addition, additional studies are needed to determine optimal equilibration periods to maximize removal of toxic chemicals using resin amendments in solid-phase sediment TIEs. As in the spiked-sediment TIEs, inconsistent elution of toxic chemicals were observed from extraction media used in solid-phase and interstitial water Phase II TIEs with the ambient samples. These procedures require additional work to ensure that chemicals responsible for toxicity are completely eluted from extraction media. Improvement of the Phase II TIE elution and solvent handling steps should facilitate better identification of chemicals responsible for toxicity, particularly in situations where toxicity is due to highly insoluble chemicals. When these considerations are satisfactorily addressed, the methods evaluated are capable of providing useful lines of evidence that can be combined to successfully identify sediment toxicants.

Benefits:

- Demonstrates that current sediment TIE procedures are effective at characterizing and identifying the cause of sediment toxicity.
- Demonstrates that the proper use of sediment TIE procedures can identify the cause of sediment toxicity and facilitate the TMDL process.
- Introduces or updates several new sediment TIE procedures that increase the current body of research.
- Identifies deficiencies in sediment TIE procedures that need to be addressed with further research.

Keywords: Sediment Toxicity Identification Evaluation (TIE), Total Maximum Daily Load (TMDL), *Eohaustorius estuarius*, *Hyalella azteca*, Ammonia, Copper, Fluoranthene, Nonylphenol, Tetrachlorobenzene

TABLE OF CONTENTS

LIST OF TABLES

LIST OF FIGURES

LIST OF ACRONYMS

AC – Alisal Creek

CS – Consolidated Slip

EDTA – Ethylenediaminetetracetic Acid

ELISA – Enzyme-Linked Immunosorbent Assay

ERL – Effect Range Low

ERM – Effect Range Median

G – Centrifugation Force

GC-MS – Gas Chromatograph / Mass Spectroscopy

HLB – Hydrophilic-Lipophilic Balance

HPLC – High Pressure Liquid Chromatography

IH – Indiana Harbor

IW – Interstitial Water

K_{ow} – Octanol-Water Partitioning Coefficient

LC50 – Median Lethal Concentration

LDD – Lowest Detectable Dose

MPSL – Marine Pollution Studies Laboratory

OC – Organic Carbon

PAH – Polycyclic Aromatic Hydrocarbon

PBO – Piperonyl Butoxide

PCB – Polychlorinated Biphenyl

PCC – Powdered Coconut Charcoal

PEL – Probable Effects Level

SD – San Diego Creek

SPE – Solid-Phase Extraction

SW – Switzer Creek

TEL – Threshold Effects Level

TIE – Toxicity Identification Evaluation

TMDL – Total Maximum Daily Load

TOC – Total Organic Carbon

TU – Toxic Unit

UN – Upper Newport Bay

Executive Summary

ES.1 TMDLs and Sediment Toxicity

Toxic pollutants discharged to surface waters may bind with particles and accumulate in the depositional areas of streams, lakes, and coastal environments. The National Sediment Quality Survey has found that 43% of nearly 20,000 stations sampled in the United States had sediment contamination levels that were probably associated with harmful effects on aquatic life or human health. Another 30% had contamination levels that were possibly associated with harmful effects (U.S. EPA 2004). Management of polluted sediments in the United States is guided by the Federal Clean Water Act (CWA), which establishes a framework of five key components: 1) standards to protect beneficial uses, 2) monitoring to determine water quality conditions, 3) assessment to determine whether standards are being attained, 4) identification of impaired waters, and 5) support for restoration of impaired waters.

Sediment toxicity tests are often used as the monitoring tool for determining whether sediments have the potential to impair aquatic life beneficial uses in contaminated areas. When toxicity tests or other biological indicators produce evidence of water quality impact from contaminated sediments, the waterbodies in which the sediments occur may be placed on the CWA §303[d] list of impaired waters. For each waterbody on the §303[d] list, states must develop Total Maximum Daily Loads (TMDLs). TMDLs must identify the cause of impairment, locate sources of the causative pollutant, and allocate loadings of the pollutant among the various sources. The goal of the TMDL is to restrict loadings so that ambient pollutant concentrations will decrease to levels that no longer contribute to impairment.

ES.2 Toxicant Identification

Identifying the cause of toxicity is the critical first step in the TMDL process. While toxicant identification can be affected by site-specific factors, there are at least three approaches that can be applied more generally to this step in the TMDL process with toxic sediments. One approach to toxicant identification is to compare chemical concentrations in the test sediment with sediment quality guideline (SQG) values developed to estimate the potential for adverse biological effects. However, these guideline values were not specifically derived for the purpose of toxicant identification, but rather for assessing the probability that adverse biological effects might occur in contaminated sediments. The relatively short list of analytes is also a limitation for the use of SQGs in toxicant identification. The second general approach to toxicant identification is to calculate the concentration of sediment-associated contaminant that is expected to dissolve into the sediment interstitial water under equilibrium conditions, assuming that interstitial water is the route of toxicant exposure. U.S. EPA has derived equilibrium partitioning sediment benchmarks for two pesticides and six metals (e.g., U.S. EPA 2003), but the approach is generally applicable to other contaminants.

This report focuses on a third approach, application of Toxicity Identification Evaluations (TIEs). The U.S. EPA and others have developed sediment TIE procedures which are designed to proceed in three phases (Ankley et al., 1991; Ho et al., 2002). Phase I manipulations characterize the classes of chemicals causing toxicity and typically differentiate between toxicity caused by organic chemicals, metals, or ammonia. Phase II TIE manipulations identify the cause of toxicity, and Phase III TIEs are designed to confirm the chemical(s) identified in Phase II. Sediment TIEs are conducted using both the aqueous matrix (sediment interstitial water) and the solid-phase matrix (whole sediment). While a growing literature has demonstrated that efficacy of TIE procedures for identifying causes of toxicity in effluents and receiving waters (Norberg-King et al., 2005), their application to sediments is less well developed. The Water Environment Research Foundation (WERF) initiated the current project to evaluate the current state of the science for accomplishing TMDLs in waterways listed as impaired for sediment toxicity, including the ability to identify the sediment-associated pollutants in need of management to restore water quality.

To evaluate and advance sediment toxicant identification in the TMDL context, a list was compiled of available methods from draft guidance documents, journal publications, draft protocols, and personal communications. These methods were developed primarily by U.S. EPA researchers, and formed the foundation for this project. In addition, a number of novel techniques designed to facilitate all phases of the TIE process for sediments were incorporated and evaluated. Both interstitial water and solid-phase TIE procedures have also been incorporated. The research to evaluate the effectiveness of toxicity identification methods for application in the TMDL process proceeded in four phases. In the first phase dose-response experiments were conducted to determine or verify toxic levels for five chemicals in marine and freshwater sediments. The five chemicals were: ammonia, copper, fluoranthene, nonylphenol, and tetrachlorobenzene. These chemicals were selected to represent classes of compounds which are common in highly polluted sediments. Ammonia was included because it is ubiquitous and its toxicity often confounds interpretation of sediment bioassays. These experiments provided critical threshold concentrations (LC50s) to guide subsequent spiking studies and also provided information on their potential toxicity in field sediments. Dose-response experiments were conducted in marine and freshwater sediments using two amphipod species, the freshwater amphipod *Hyalella azteca* and the estuarine amphipod *Eohaustorius estuarius*. Both are standard sediment testing species commonly used in sediment quality assessments (U.S. EPA 1994c; U.S. EPA 2000). All spiking experiments were conducted with formulated sediment developed as part of this project.

ES.3 Single Chemical Evaluations

In the second phase of the project, formulated sediment was spiked with each of the five model chemicals and these single-chemical-spiked sediments were subjected to solid-phase and interstitial water TIE procedures. This phase evaluated the methods under controlled conditions and identified techniques that were most diagnostic for the specific classes of compounds. These experiments emphasized Phase I (chemical characterization) and II (chemical identification) TIEs procedures. For interstitial water Phase II procedures, techniques to elute chemicals from solid-phase extraction columns were evaluated. Because of the limited methodology for identifying individual compounds in solid-phase Phase II TIEs, techniques to elute chemicals from the carbonaceous resin Ambersorb 563 were developed during this phase of this project. The current results demonstrated that these procedures can successfully characterize the class of chemicals responsible for toxicity and identify the causes of toxicity. There were some inconsistencies between the solid-phase and interstitial water techniques, particularly in the ability to recover toxic concentrations of specific contaminants from resins and extraction media used in the sediment and interstitial water TIEs. For example, the technique was largely unsuccessful in recovering toxic concentrations of copper in experiments with the marine amphipod *E. estuarius*. This was due in part to this species' relative insensitivity to copper, and to an apparent lack of cupric ion bioavailability when acid eluates of copper were spiked into seawater. The primary conclusion from the single chemical TIEs was that the interstitial water and solid-phase TIE procedures often provide complementary results, and TIEs are more successful when both approaches are used.

ES.4 Mixture Evaluations

In the third phase of the project, formulated sediment with the above chemicals in various mixtures was spiked to evaluate methods to resolve toxicity due to multiple chemicals. TIEs were conducted with both interstitial water and solid-phase sediment matrices, using all appropriate techniques identified in the earlier phases. Phase 3 of the project identified the TIE methods most useful for resolving toxicity due to multiple chemicals, including ammonia. These experiments corroborated results of the Phase 2 experiments, and demonstrated the utility of combining solid-phase and interstitial water TIEs. For example, formulated sediment was spiked with mixtures of copper and fluoranthene at nominal concentrations approximately four times their respective LC50s. In addition, this mixture was spiked with sublethal concentrations of nonylphenol, ammonia and tetrachlorobenzene. The results showed reduced toxicity to the marine amphipod *E. estuarius* after additions of resins to bind metals and organics. In addition, toxicity was observed in water spiked with solvent extracts of the two resin amendments and subsequent chemical analyses confirmed that these contained toxic concentrations of copper and fluoranthene. When combined with results of similar TIE experiments conducted on interstitial water extracted from this sediment, the weight-of-evidence confirmed that these methods could both characterize and identify causes of toxicity in chemical mixtures. Similar marine experiments were conducted with mixtures containing toxic concentrations of copper and ammonia, and nonylphenol and ammonia. All of the mixture experiments were then repeated with freshwater sediments tested with *H. azteca*. In mixtures containing toxic concentrations of ammonia and copper or ammonia and nonylphenol, zeolite was particularly useful at reducing unionized ammonia from interstitial water. This allowed for the use of other techniques to remove the remaining copper or nonylphenol.

As before, the primary conclusion was that the ability to identify the causes of toxicity in spiked sediments depended on various lines of evidence derived from both interstitial water and solid-phase TIEs and that success depended on applying a number of corroborating techniques in each TIE. These experiments also illustrated that Phase II chemical elution procedures require additional refinement. For example, the ability to extract some chemicals from either the resins used in solid-phase TIEs or the solid-phase extraction columns used in interstitial water TIEs was inconsistent. It is recommended that additional research be conducted using different solvents and extraction media to further develop these procedures.

ES.5 Ambient Sample Evaluations

The final phase of this project was designed to validate the techniques evaluated in the previous phases by applying these methods to TIEs of ambient sediment samples. The researchers collected marine and freshwater sediments from six locations, three from freshwater habitats and three from marine habitats. These sediments were selected to represent different input scenarios: agricultural runoff, urban and industrial runoff, and industrial effluents; and were selected in consultation with the members of the Project Scientific Committee. Five of the six samples were from California. Marine samples collected in southern California were from Switzer Creek in San Diego Bay, Upper Newport Bay in Newport Beach, and Consolidated Slip in Los Angeles Harbor. The two freshwater samples from California were San Diego Creek in Newport Beach, and Alisal Slough in the City of Salinas, in central California. The final freshwater sediment was from Indiana Harbor. Interstitial water and solid-phase toxicity tests were conducted on each sample, applying appropriate TIE procedures depending on the magnitude of toxicity exhibited by the different sediment matrices. The results suggest that these methods were successful at differentiating between sediment toxicity caused by ammonia, metals, and organic chemicals in both marine and freshwater samples. In the majority of cases, treatment of solid-phase samples with either Ambersorb 563, SIR-300 or powdered coconut charcoal reduced toxicity. In the one sample containing toxic concentrations of unionized ammonia (Indiana Harbor), addition of zeolite was effective at reducing this constituent, prior to application of additional Phase I treatments. Where interstitial water toxicity was observed in these samples, Phase I treatments were usually effective at characterizing toxicity due to metals, organics or ammonia. The Phase I TIE methods were particularly useful when the results of solid-phase and interstitial water procedures were considered together. The results demonstrated that the Phase I TIE treatments were less effective in highly toxic samples. Almost all of the samples required some degree of dilution before toxicity reduction was observed using these procedures.

Mixed results were observed using the Phase II TIE methods with the ambient samples. As part of the Phase II TIE solid-phase procedures, the resin amendments (SIR-300 or Ambersorb) were recovered from the sediments and eluted with either acid (SIR-300) or acetone (Ambersorb) to determine whether toxic concentrations of cationic metals or organic chemicals were returned to clean water spiked with the eluates. In the majority of TIEs of the ambient samples, the solvent eluates of media used to extract organic chemicals from sediment or interstitial waters were toxic when spiked into clean water. Chemical analyses indicated that the ambient sediment samples contained high concentrations of relatively insoluble organic compounds (e.g., pyrethroid pesticides, PAHs, PCBs), and the results indicated that hydrophobic compounds are difficult to elute, and subject to loss through sorption and dilution at different steps in the TIE and analytical chemistry processes. Analyses of the original Ambersorb acetone eluates and the HPLC fractions from these eluates showed very little detection of organic chemicals. After consultation with the chemists, it was determined that the eluates and HPLC fractions had been subjected to additional cleanup steps prior to the gas chromatography analysis, and this resulted in significant sample dilution. When the chemists directly injected the acetone extracts and HPLC fractions, and reanalyzed the samples using GC-MS, toxic concentrations of pyrethroid pesticides were detected in four of the ambient samples, with the results of a fifth sample pending. This illustrates the importance of constant communication between toxicologists and chemists involved in TIEs. Because TIEs involve an iterative process where results of chemical analysis from initial steps direct procedures in subsequent TIE steps, direct communication and rapid turn-around times are necessary. Modifications to the current Phase II procedures to address these deficiencies are suggested in the section of the report that discusses future research needs.

The primary approach to Phase III (confirmation) TIEs in the current study was to compare chemical concentrations in sediments and interstitial waters to known effect levels derived from dose-response experiments (i.e., species sensitivity approach). When these data were not available, concentrations were compared to empirically derived sediment quality guideline values, or model-based values. Many of these samples were highly contaminated by complex chemical mixtures. Many of the samples contained individual chemicals that exceeded toxicity thresholds based on laboratory dose-response experiments, but some also contained mixtures of chemicals for which no dose-response information existed. While chemicals in these samples exceeded empirically derived sediment quality guideline values and toxicity thresholds based on model-derived values, the results illustrate the difficulty of confirming toxicity that is most likely due to the interaction of a number of organic constituents and unionized ammonia. This difficulty is caused by two interrelated considerations. First, when Phase II TIE evidence suggests multiple chemicals are present, current methods are insufficient to remove any single compound on its own, so that the effects of the other chemicals can be evaluated in its absence. Secondly, it is currently not possible to measure quantitatively the proportion of the total amount of a chemical in the sediment that is removed by the resin amendments. It is not possible, therefore, to add back an equal amount to clean sediment to confirm identification, as can be done with water-only TIEs. These difficulties with solid-phase confirmations steps underscore the utility of using both solid-phase and interstitial water matrices in the TIE process.

ES.6 Future Refinements

In this report the utility of existing and more recently developed sediment TIE procedures were evaluated for characterizing, identifying, and confirming the cause(s) of sediment toxicity using spiked-laboratory and ambient samples. The results indicate that sediment TIE methods are sufficiently developed to characterize toxicity due to general classes of chemicals such as cationic metals, organic chemicals, and ammonia. Methods to improve extraction and elution of metal and organic chemicals in highly toxic sediments require further refinement in order to allow identification of specific chemicals responsible for toxicity as part of the Phase II TIE process. These refinements include determining the appropriate masses of solid-phase extraction media to allow complete removal of toxic chemicals from sediments and interstitial waters. In addition, additional studies should be designed to determine optimal equilibration periods to maximize removal of toxic chemicals using resin amendments in solid-phase sediment TIEs. The current project indicated inconsistent elution of toxic chemicals from extraction media used in solid-phase and interstitial water Phase II TIEs and these procedures require additional work to ensure that chemicals responsible for toxicity are completely eluted from extraction media. Similarly, further refinements are needed in the selection and handling of solvents used to elute the resin amendments and columns that remove toxicants from the sediment samples. These solvent refinements are necessary to optimize chemical analysis for the identification of hydrophobic and highly toxic chemicals that must be measured at low concentrations to assess their contributions to sample toxicity. Such low concentrations can be decreased or lost in solvent exchange and handling processes between the elution step and final chemical identification. Improvement of the Phase II TIE elution and solvent handling steps should facilitate better identification of chemicals responsible for toxicity using gas chromatography and other analytical techniques, particularly in situations where toxicity is due to highly insoluble chemicals.

Each TIE, and especially each sediment TIE, is best considered as a unique experiment, but one that will more likely succeed if experienced chemists and toxicologists apply the best available techniques. The results of this study indicate that the techniques evaluated, when carefully applied, can provide solid lines of evidence that, taken together, are capable of identifying chemicals causing toxicity in field-collected sediments. The lines of evidence that appear particularly useful are those drawn from sorption and elution of solid-phase resin amendments (SIR-300 and Ambersorb), and extraction and elution of interstitial water constituents using solid-phase extraction columns. Refinements or modifications are likely to be necessary with each individual sediment TIE, and care should be taken to keep the following considerations in mind:

1) It is difficult to identify less toxic (but contributing) compounds in highly toxic sediments, as their toxicity may be lost when sediments are diluted to allow resolution of responses to TIE manipulations.
2) Efforts should be made to optimize the efficiency of contaminant sorption to and solvent elution from the resin amendments and columns used to extract toxicants from sediment and interstitial water.
3) Toxicologists and chemists must closely coordinate the selection and processing of solvents used to elute contaminants from resins and columns. This is of particular importance for highly insoluble, highly toxic compounds that are active at low concentrations that can easily be lost through sorption and exchange. Ideally, the solvents used for extraction should be suitable for injection into analytical instruments with minimal transfer, exchange, or storage time.

When these considerations are satisfactorily addressed, the methods evaluated are capable of providing useful lines of evidence that can be combined to successfully identify sediment toxicants.

ES.7 TIEs in the TMDL Process

As with TIEs, each sediment TMDL is best considered as a unique study that can benefit from previous experience and established tools. While the toxicant identification step follows a general approach, as described above, many other aspects of TMDL development are highly site-specific. Unique combinations of soils, hydrology, and human activities add to the complexity of biogeochemical processes affecting contaminant sorption to sediments, persistence in suspended and bedded sediments, and bioavailability. Contaminants may enter the system from both point and non-point sources, and may enter either as dissolved chemicals that bind to sediments in receiving waters, or as chemicals already bound to sediments through landscape processes such as transport across urban or agricultural surfaces. In-stream contaminant transport processes are complicated both by particle sorption and desorption kinetics, and by the dynamics of sediment deposition and resuspension during tidal cycles, storm events, and seasonal flow stages.

Most, if not all, of the processes mentioned above are driven by site-specific characteristics that must be evaluated on a case by case basis during TMDL development. Once the contaminant of concern is identified, methods exist to estimate target concentrations and source contributions, but these methods have not been gathered, evaluated, or presented in the context of sediment toxicity TMDLs. There are rich literatures addressing bioavailability, equilibrium partitioning, watershed and hydrologic modeling, sediment transport, and effluent characterization, but there are few if any published studies that combine this information into a framework for a general approach to sediment toxicity TMDLs. This report is the most thorough review and evaluation to date of the methods necessary to conduct the critical first step of toxicant identification. Expert scientific guidance is needed to complete the remaining steps of setting target concentrations and determining the load restrictions necessary to achieve standards that protect aquatic life beneficial uses in waterways impaired by sediment toxicity.

CHAPTER 1.0

INTRODUCTION

Toxic pollutants discharged to surface waters tend to bind with sediment particles and accumulate in the depositional areas of streams, lakes, and coastal environments. The National Sediment Quality Survey has found that 43% of nearly 20,000 stations sampled in the United States had sediment contamination levels that were probably associated with harmful effects on aquatic life or human health. Another 30% had contamination levels that were possibly associated with harmful effects (U.S. EPA, 2004). Management of polluted sediments in the U.S. is guided by the Federal Clean Water Act (CWA), which requires the establishment of standards to protect water quality. Because of the complex geochemical processes that govern bioavailability of sediment-associated contaminants, few numeric sediment quality standards have been promulgated. In place of numeric standards, many states use narrative standards to determine whether sediment contaminants are impacting beneficial uses of surface waters. Narrative standards are based on the CWA requirement that waters be free of toxic substances in toxic amounts. Sediment toxicity tests are often used to determine whether sediments have the potential to impair aquatic life in contaminated areas.

When toxicity tests or other biological indicators produce evidence of water quality impairment from contaminated sediments, the waterbodies in which the sediments occur may be placed on the CWA §303[d] list of impaired waters. For each waterbody on the §303[d] list, states must develop Total Maximum Daily Loads (TMDLs). TMDLs must identify the cause of impairment, locate all sources of the causative pollutant, and allocate loadings of the pollutant among the various sources. The goal of the TMDL is to restrict loadings so that ambient pollution concentrations will decrease to levels that no longer contribute to impairment.

The 1998 CWA §303[d] list contained about 20,000 impaired waterbodies nationwide, which needed about 36,000 TMDLs to be developed for the applicable stressors. Based on the TMDL tracking system with the 1998 data, only 32 impaired waterbodies were specifically identified as impaired by contaminated sediments. However, about 21% of all TMDLs were for pollutants that are often found in contaminated sediments (e.g., mercury, pesticides, PCBs and other industrial organics; USEPA 2004).

Developing TMDLs for waterbodies listed as impaired by sediment toxicity involves a number of scientific challenges. As with waterbodies listed for water toxicity, "toxicity loads" cannot be allocated, so water quality improvement depends on identifying the pollutant(s) that cause the observed toxicity, so that loadings of the pollutant(s) can be managed. Identifying the cause of toxicity is a crucial step in the TMDL process.

With sediment toxicity, there are additional considerations due to the complex biogeochemical processes affecting contaminant sorption to sediments, persistence in suspended and bedded sediments, and bioavailability. Benthic, epibenthic, and water column organisms can be exposed through interstitial water, flux to overlying water, and/or ingestion of sediment particles. In some situations, the sediment itself is considered the habitat of concern, because many pollution-sensitive organisms live in the sediment and play key roles in both benthic and water column ecosystems. In other cases, the sediment might be considered as a source of contaminants to a water column habitat of concern. Both scenarios are likely to apply in most cases. Another consideration for sediment toxicity TMDLs is that contaminants may enter the system from both point and non-point sources, and may enter either as dissolved chemicals that bind to sediments in receiving waters, or as chemicals already bound to sediments through landscape processes such as transport across urban or agricultural surfaces. Finally, instream contaminant transport processes are complicated both by particle sorption and desorption kinetics, and by the dynamics of sediment deposition and resuspension during tidal cycles, storm events, and seasonal flow stages. Most, if not all, of these processes are driven by site-specific characteristics that must be evaluated on a case by case basis during TMDL development.

While toxicant identification can be affected by site-specific factors, there are at least three approaches that can be applied more generally to this critical first step in the TMDL process. One approach to toxicant identification is to compare chemical concentrations in the test sediment with sediment quality guideline values developed to estimate the potential for adverse biological effects. Generally accepted sediment quality guideline (SQG) values have been developed for about 30 chemicals, including: four aroclors and total PCBs in marine and freshwater (MacDonald et al., 2000a); 13 polycyclic aromatic hydrocarbons (PAHs; Swartz, 1999); nine metals, 13 PAHs, two pesticides, and total PCBs in marine waters (Long et al., 1995; Long and Macdonald, 1998); and a similar list of compounds in freshwater (MacDonald et al., 2000b). However, these guideline values were not derived for the purpose of toxicant identification, but rather for assessing the probability that adverse biological effects might occur in contaminated sediments. The relatively short list of analytes is also a limitation for the use of SQGs in toxicant identification.

The second general approach to toxicant identification is to calculate the concentration of sediment-associated contaminant that is expected to dissolve into the sediment interstitial water under equilibrium conditions, assuming that interstitial water is the route of toxicant exposure. Equilibrium partitioning sediment benchmarks (ESB) are derived by multiplying the chemical's octanol/water partitioning coefficient, K_{ow}, by the concentration of the chemical which would cause toxicity in water, yielding the concentration in solid-phase sediment (normalized to organic carbon) that would be expected to cause toxicity. U.S. EPA has derived equilibrium partitioning sediment benchmarks for two pesticides and six metals (e.g., U.S. EPA, 2003), but the approach is generally applicable to other contaminants.

The third approach to toxicant identification that can be applied to sediment toxicity TMDLs is the Toxicity Identification Evaluation (TIE). TIEs involve conducting an initial test to determine the presence and magnitude of toxicity, followed by sample manipulation to decrease, transform, or change the bioavailable fractions of sample contaminants, so that their contributions to sample toxicity can be assessed (e.g., Ankley et al., 1991; U.S. EPA, in press). TIEs generally involve a weight-of-evidence approach via process of elimination, and can include application of SQG and ESB values to screen potential toxicants based on their measured concentrations in baseline samples and TIE treatments.

This report focuses on the TIE approach to identifying pollutants in need of management through the TMDL process for waterbodies impaired by sediment toxicity. The goal of this study is to evaluate existing and emerging sediment TIE methods, and to determine which techniques can provide credible lines of evidence leading to toxicant identification. The research team evaluated the techniques using sediments spiked with individual chemicals and mixtures, and with field sediments from contaminated waterbodies. The ultimate goal of this research was to evaluate whether available TIE techniques are sufficient to address the critical first step in the TMDL process for toxic sediments, which is to identify chemical(s) responsible for toxicity.

TIEs are designed to proceed in three phases. Phase I manipulations characterize the classes of chemicals causing toxicity and typically differentiate between toxicity caused by organic chemicals, metals, or ammonia. Phase II TIE manipulations identify the individual chemicals causing toxicity, and Phase III TIEs are designed to confirm the Phase II chemical identification. TIE method development for sediments has followed two tracks: 1) identification of toxicants in sediment interstitial water (e.g., Ankley et al., 1991), and 2) identification of toxicants in solid-phase sediment (e.g., Burgess et al., 2000; Lebo et al., 2000). Interstitial water is thought to be the primary medium through which organisms are exposure to many chemicals, especially lower K_{ow} compounds (Ho et al., 2002), and interstitial water TIEs have the advantage of using an aqueous medium that is more amenable to standardized effluent TIE techniques. Solid-phase sediment TIEs have the advantage of maintaining more realistic conditions for oxidation state, pH, and other factors affecting contaminant partitioning, but the medium is less amenable to TIE manipulations. Fewer TIE techniques have been available for solid-phase TIE.

This study was designed to evaluate the effectiveness of existing sediment TIE procedures, to determine their ability to support the contaminant identification step in TMDLs for waterbodies listed as impaired by sediment toxicity. To evaluate and advance sediment toxicant identification, methods available in draft guidance documents, journal publications, draft protocols, and personal communications were compiled. These methods formed the foundation for this project, and the methods presented here are based largely on those developed by the U.S. EPA researchers who authored many of these documents. In addition, a number of emerging techniques designed to facilitate all phases of the sediment TIE process were incorporated and evaluated. Both interstitial water and solid-phase TIE procedures have been incorporated in this project. The research to evaluate the effectiveness of toxicity identification methods for application in the TMDL process proceeded in four phases. In the first phase dose-response experiments were conducted to determine or verify toxic levels for five chemicals in marine and freshwater sediments. The five chemicals were: copper; fluoranthene, tetrachlorobenzene, nonylphenol, and ammonia. These chemicals were selected to represent a range of chemical classes that are commonly measured in highly polluted sediments. Ammonia was included because it is fairly ubiquitous and its toxicity often confounds interpretation of sediment bioassays. These experiments provided critical threshold concentrations to guide subsequent spiking studies and also provided information on their potential toxicity in field sediments. Dose-response experiments were conducted in marine and freshwater sediments using two species of infaunal amphipods, the freshwater epibenthic scavenger *Hyalella azteca* and the estuarine free-burrowing detritivore *Eohaustorius estuarius*. Both are standard sediment testing species commonly used in sediment quality assessments (U.S. EPA, 1994c; U.S. EPA, 2000).

In the second phase of the project, formulated sediment was spiked with each of the five model chemicals and these single chemical spiked sediments were subjected to solid-phase and interstitial water TIE procedures. In this phase the methods were evaluated under controlled conditions and identified techniques that were most effective or diagnostic for the specific classes of compounds. These experiments emphasized Phase I (chemical characterization) and II (chemical identification) TIEs procedures. For interstitial water Phase II procedures, techniques were evaluated to improve elution of chemicals from solid-phase extraction columns. Because of the limited methodology for identifying individual compounds in solid-phase Phase II TIEs, techniques were developed to elute chemicals from the carbonaceous resin Ambersorb 563 (Rohm and Haas, Spring House, PA, U.S.) during the second phase of this project.

In the third phase of the project, formulated sediment was spiked with the above chemicals in various mixtures to evaluate methods to resolve toxicity due to multiple chemicals. In these experiments, sediments were spiked with toxic concentrations of ammonia and copper, and with toxic concentrations of ammonia and fluoranthene. In subsequent experiments all five chemicals were mixed to provide mixtures of greater complexity. TIEs were then conducted using both interstitial water and solid-phase sediment matrices using all appropriate techniques identified in the earlier phases. Phase 3 of the project identified the TIE methods most useful for resolving toxicity due to multiple chemicals, including ammonia.

The final phase of this project was designed to field test the techniques evaluated in the previous phases by applying these methods to TIEs of ambient sediment samples. Marine and freshwater sediments were collected from six locations, three from freshwater habitats and three from marine habitats. These sediments were selected to represent different input scenarios: agricultural runoff, urban and industrial runoff, and industrial effluents, and were selected in consultation with the WERF Project Subcommittee that provided scientific advice for the study. Interstitial water and solid-phase toxicity tests were then conducted on each ambient sediment sample, applying appropriate TIE procedures depending on the magnitude of toxicity exhibited by the different sediment matrices.

In this report, the utility of existing and more recently developed sediment TIE procedures for characterizing, identifying, and confirming the cause(s) of sediment toxicity were evaluated. These results were considered in the context of TMDL development for waterbodies impaired by sediment toxicity, and the most promising methods for sediment toxicant identification were emphasized. The conclusion identifies areas of future research.

CHAPTER 2.0

MATERIALS AND METHODS

2.1 Sediment TIE Approach

The primary objectives of this study were to evaluate current U.S. EPA sediment TIE methods, develop additional procedures where needed, and determine which methods are most effective for identifying individual chemicals responsible for sediment toxicity. A list of the evaluated solid-phase and interstitial water methods is summarized in Table 2-1. This table includes a brief description of the purpose of the method and provides citations for each method. The complete procedures for each TIE method are described in detail in Appendix A. A list has been compiled of the most effective procedures, based on the results of the experiments evaluating sediments spiked with single and multiple chemicals, as well as the results of six ambient sample TIEs. These treatments are structured in a decision matrix (Figure 2-1) that provides guidelines for Phase I and Phase II of the sediment TIE process. Note that several procedures published by U.S. EPA for application in sediment TIEs were not included in the current study. These included addition of Acid-Volatile Sulfide (AVS) to manipulate bioavailability of cationic metals, addition of the green alga *Ulva sp.* to remove unionized ammonia, and use of semi-permeable membrane devices as a method to evaluate bioavailability of organic chemicals extracted from sediments. Methods and results using these procedures are described in detail in a recent sediment TIE guidance document (U.S. EPA, in press).

After sample toxicity has been demonstrated, the decision matrix guides the TIE process. TIEs proceed with an initial assessment of the magnitude of toxicity using solid-phase and interstitial water dilution series tests. For example, solid-phase dilutions may be necessary when complete mortality is observed in the original sample, because this allows better resolution between baseline and TIE treatments. Once the magnitudes of toxicity are determined, a single sediment concentration is selected for the solid-phase TIE, and a dilution series is selected for the interstitial water TIE. In this approach, solid-phase TIEs are generally conducted on a single concentration of sediment for efficiency, but could be conducted on multiple concentrations if resources allow. The solid-phase concentration is chosen to provide the best resolution among the TIE treatments. Interstitial water TIEs are conducted with a dilution series that brackets the LC50 from the initial test. In situations where one sediment matrix produces insufficient toxicity to allow resolution of differences between the TIE treatment and baseline, the TIE proceeds using the matrix with the highest magnitude of toxicity.

In the current study ambient solid-phase sediment samples were diluted with either amphipod collection site sediment (*E. estuarius*) or formulated sediment (*H. azteca*). The concentrations were based on the results of the original test and any other information available for the sample. The sediment concentrations in the ambient sample TIEs conducted in this study ranged from 1% to 100%. Prior to the initiation of solid-phase TIE treatments, the diluted sample was prepared by combining wet weights of sample and dilution sediment in a mixing jar ranging from 500 to 2000 mL depending on the volume needed. The sediment mixture was rolled for 24 hours on a sediment roller apparatus.

In all experiments, interstitial water was extracted using a refrigerated centrifuge (2500 *g*, 4 °C), and diluted with clean control water prepared at the appropriate salinity. In the current study, samples tested with *E. estuarius* were diluted with 20‰ water prepared by combining 0.45-µm filtered natural seawater with distilled water. Samples tested with *H. azteca* were diluted with laboratory well water. Diluted samples were used immediately in the TIE procedures.

Diluting the sediment for a solid-phase TIE can significantly affect the chemistry of the sample and therefore influence interpretation of the TIE results. The role of contaminants that are marginally toxic in an undiluted sample could be reduced through dilution, and thus be overlooked in TIEs conducted at lower concentrations. To address this issue, conducting solid-phase TIE treatments on both diluted and undiluted sediment is recommended. Solvent elution of carbonaceous resins and acid elution of cation resins are particularly informative for undiluted solid-phase samples, and data from these treatments can be combined with information from lower dilutions to add to the weight-of-evidence in a TIE. The researchers further recommend conducting interstitial water TIEs using a dilution series. This allows for the calculation of toxic units (TUs), and greatly aids in the interpretation of TIE data. Regardless of what lower dilutions are used, an undiluted sample should be included in the interstitial water TIE.

2.2 Sediment and Interstitial Water Methods

The TIE treatments evaluated in this study were compiled from existing U.S. EPA guidance documents and protocols, and from a review of the recent TIE literature. Every effort was made to evaluate a complete list of solid-phase and interstitial water sediment TIE methods, but there were inevitable omissions. The omitted methods will be discussed where appropriate in the context of the overall sediment TIE evaluation. A complete list of evaluated methods and citations is provided in Table 2-1, and detailed procedures are listed in Appendix A. The decision matrix used in the current study (Figure 2-1) contains the list of methods that proved to be most effective at reducing the toxicity of target chemical classes.

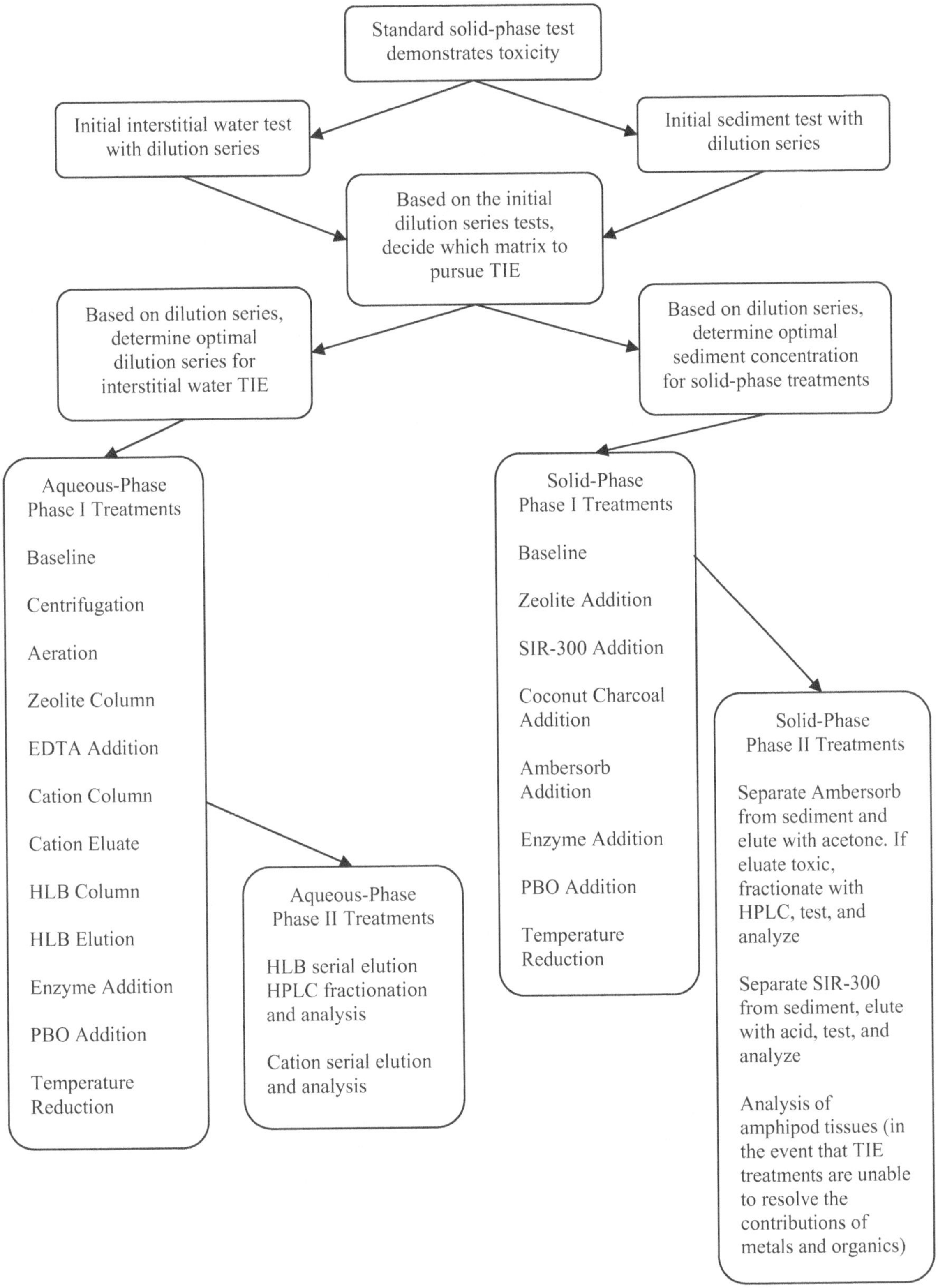

Figure 2-1. Sediment TIE Decision Matrix.

Table 2-1. Summary of Solid-phase and Interstitial Water TIE Methods. Complete descriptions of the methods are contained in Appendix A.

Solid-Phase Method	Description	Citations
Zeolite (SIR-600)	Addition of 10-20% zeolite to sediment reduces interstitial and overlying water ammonia	Besser et al., 1998; Burgess et al., 2003
Chelating Resin (SIR-300)	Addition of 10% SIR-300 chelates heavy metal ions and reduces metal bioavailability	Burgess et al., 2000
Coconut Charcoal (PCC)	Addition of 10-15% PCC to sediment reduces toxicity caused by organic contaminants	Lebo et al., 1999, 2003; Ho et al., 2002, 2004
Ambersorb 563	Addition of 10% Ambersorb to sediment reduces toxicity caused by organic contaminants	Kosian et al., 1999; Lebo et al., 1999; West et al., 2001
Carboxylesterase	Addition of enzyme to overlying water reduces toxicity caused by pyrethroid pesticides by breaking down compounds into non-toxic forms	Wheelock et al., 2004, 2006
Piperonyl Butoxide	Addition of PBO to overlying water can reduce toxicity caused by organophosphate pesticides, and increase toxicity caused by pyrethroid pesticides and DDT	U.S. EPA, 1991; Kakko et al., 2000
Temperature Reduction	Testing warm water organisms at colder temperatures can increase the toxicity of pyrethroids and DDT	Ware 1989

Interstitial Water Method	Description	Citations
Aeration	The sample is aerated for one hour to determine if toxicity is caused by volatile compounds or surfactants	U.S. EPA, 1991, 1993, 1996
Filtration or Centrifugation	Reduces toxicity that is particle-related. Also used as a pretreatment step for the column treatments	U.S. EPA, 1991, 1996
EDTA	Organic chelating agent that preferentially binds with divalent metals, such as copper, nickel, lead, zinc, cadmium, mercury, and other transition metals to form non-toxic complexes	U.S. EPA, 1991, 1996
pH Adjustment and Volatilization	Reduces ammonia in sample by converting it to the unionized fraction through pH adjustment and volatilizing it by stirring	U.S. EPA, 1991, 1993; Burgess et al., 2003
Graduated pH	Determines if pH-dependent toxicants are responsible for the observed toxicity. The toxicity of ammonia, sulfide and some metals change with pH	U.S. EPA 1991, 1993, 1996
Zeolite Column Solid-Phase Extraction	Reduces toxicity caused by ammonia	U.S. EPA, 1991; Burgess et al., 2004
Cation Column Solid-Phase Extraction and Elution	Removes metals from the sample. Column can be eluted with 1N hydrochloric acid (HCl) and resulting eluate tested to determine if substances removed by the column were toxic	U.S. EPA, 1993, 1996; Burgess et al., 1997
Non-Polar Organic Column Solid-Phase Extraction and Elution	Removes non-polar organic compounds. Column can be eluted with solvent to determine if substances removed by the column were toxic	U.S. EPA, 1991,1993, 1996
Carboxylesterase	Addition of enzyme to sample reduces toxicity caused by pyrethroid pesticides by breaking down compounds into non-toxic forms	Wheelock et al., 2004, 2006
Piperonyl Butoxide	Addition of PBO can reduce toxicity caused by organophosphate pesticides, and increase toxicity caused by pyrethroid pesticides and DDT	U.S. EPA, 1991; Kakko et al., 2000
Temperature Reduction	Testing warm water organisms at colder temperatures can increase the toxicity of pyrethroids and DDT	Ware, 1989

Sediment and interstitial water toxicity in this study were evaluated using the estuarine amphipod *Eohaustorius estuarius* or the freshwater amphipod *Hyalella azteca*. *H. azteca* solid-phase exposures followed U.S. EPA (2000), and *E. estuarius* exposures followed U.S. EPA (1994c), with the following modifications. Five replicate 250-mL beakers containing 50 mL sediment and 200 mL overlying water were used for each chemical concentration. The exposures were conducted under static conditions for 10 days with slow aeration. Freshwater tests used ten *H. azteca* individuals per beaker and marine tests used five *E. estuarius* individuals per beaker. Solid-phase tests with *H. azteca* were fed one mL YCT per beaker every other day. Interstitial water exposures were conducted in 20-mL scintillation vials containing 10 mL of water. *E. estuarius* tests were conducted with five replicates, each containing one amphipod, and *H. azteca* tests were conducted with three replicates, each containing five amphipods. Interstitial water tests were conducted for up to ten days depending on the results. Although the interstitial water TIEs were conducted for ten days, daily survival was quantified to determine the results that provided the best resolution among the TIE treatments. All *E. estuarius* were obtained from Northwestern Aquatic Sciences (Newport, OR) and were 3 to 5 mm. All *H. azteca* were obtained from Chesapeake Cultures (Hayes, VA) and were 7 to 14 days old.

2.3 Formulated Sediment and Spiking Methods

Formulated sediment was used in all spiking experiments. This sediment was created using equal parts reference site sediment from the Salinas River (Monterey County, CA), and clean, kiln-dried sand (#60, RMC Pacific Materials, Monterey, CA, U.S.). The sediment was amended with 0.75% organic peat moss (Uni-Gro, Chino, CA, U.S.). One kilogram (dry weight) of formulated sediment was prepared by combining 496.25g reference sediment, 496.25g sand, and 7.50g peat with 350 mL water. Laboratory well water was used for freshwater tests with *H. azteca*, and laboratory seawater (adjusted to 20‰ with distilled water) was used for estuarine tests with *E. estuarius*. Three separate batches of formulated sediment were used during the study. Before use, each batch was tested with amphipods and then characterized chemically. Trace organic analysis demonstrated non-detects for most contaminants (Table C-1). Contaminants that were detected were well below the tolerance threshold of the test organisms. The average grain size of the formulated sediment was 61.18% sand and 38.82% fines, and the average total organic carbon (TOC) was 0.75% (Table C-1).

Methods used to spike copper and ammonia into formulated sediment differed from those used to spike fluoranthene, nonylphenol and tetrachlorobenzene. Ammonia and copper stock solutions were spiked directly into the water used to prepare the formulated sediment. These sediments were then rolled on a Wheaton roller apparatus (Wheaton Instruments, Millville, NJ, U.S., Model 348940-48) for the first 24 hours of the equilibration period. Organic chemical stock solutions were prepared in acetone and placed into 2000-mL glass jars where the solvent was allowed to volatilize (Ditsworth et al., 1990). The jars were rolled under a fume hood with their lids off to facilitate volatilization. After volatilization, formulated sediment was prepared in the containers and rolled for 24 hours at 15 °C. After initial rolling for all five test chemicals, the jars were transferred to a refrigerator for the remaining equilibration period at a temperature of 4 °C. Equilibration times were as follows: ammonia, 48 hours; copper, 14 days (Simpson et al., 2004); fluoranthene and tetrachlorobenzene, 28 days (Cole et al., 2000); and nonylphenol, 7 days (Fay et al., 2000).

Concentrations for the single chemical spikes were chosen based on median lethal concentrations (LC50s) calculated from the results of range finding tests (Table 2-2). Once the LC50 was calculated, single-chemical TIE sediments were spiked at a range of 1.2 to 4.1 toxic units. In the multiple chemical TIEs, pairs of dominant chemicals were spiked using the same concentrations used in the single chemical TIEs. The additional chemicals were spiked at approximately one-quarter TU (Table 2-2).

Concentrations of spiked chemicals were measured in selected sediment and interstitial water samples. Ammonia concentrations were measured in sediment overlying water and interstitial water using a spectrophotometric salicylate method (Hach Company, Loveland, CO). Copper was measured in sediment using U.S. EPA Method 1640 (1995) for seawater, U.S. EPA Method 200.8 (1994a) for freshwater, or U.S. EPA Method 6020 (1990) for sediments. Fluoranthene and tetrachlorobenzene were measured using U.S. EPA Method 8270 (1994b).

The following classes of chemicals were also measured in the ambient samples: metals (U.S. EPA Methods 200.8, 1640, and 6020), pyrethroid pesticides (Method 1660; U.S. EPA, 1993c), organochlorine pesticides (Method 8081; U.S. EPA, 1994b), organophosphate pesticides (Method 8141; U.S. EPA, 1994b), polychlorinated biphenyls (Method 8082; U.S. EPA, 1994b), and polycyclic aromatic hydrocarbons (Method 8270; U.S. EPA, 1994b). Detection limits for all chemicals are listed in Appendix C.

Interstitial water sampled from some of the ambient sediments was also analyzed for the organophosphate pesticides chlorpyrifos and diazinon with enzyme-linked immunosorbent assays (ELISA). Measurements were compared to a five-point standard curve. The lowest detectable dose (LDD) was calculated from analysis of laboratory standards according to the manufacturer's methodology (Strategic Diagnostic, Newark, DE). The LDD was the amount of pesticide required to achieve a ratio of 85% between the mean absorbance of the standard and the mean absorbance of a negative control or laboratory control water (Sullivan and Goh, 2000). Absorbance is inversely proportional to concentration. The diazinon LDD was 30 ng/L, and the chlorpyrifos LLD was 50 ng/L. Reporting limits were two times the LLD values. External standards and sample duplicates were measured. Recovery of external standards ranged from 92% -120% for chlorpyrifos and 115% -130% for diazinon. Duplicates had coefficients of variation less than 10%.

Table 2-2. Summary of Dose-response Statistics Calculated from Nominal Sediment Concentrations, and Target Spiking Concentrations for Single and Multiple Chemical Spikes. All units are mg/kg.

	Summary Statistics			Target Spiking Concentrations			
Chemical	LC50[1]	NOEC[2]	LOEC[3]	Single Chemical	Copper/ Fluoranthene	Copper/ Ammonia	Nonylphenol/ Ammonia
E. estuarius							
Ammonia	318	180	320	600	75	600	600
Copper	584	560	1000	2000	2000	2000	200
Fluoranthene	85.3	56	100	100	100	15	15
Nonylphenol	61.3	50	100	250	15	15	250
Tetrachlorobenzene	195	100	250	800	50	50	50
H. azteca							
Ammonia	800	560	1000	1500	200	1500	1500
Copper	260	32	100	1000	1000	1000	50
Fluoranthene	90.6	56	100	200	200	30	30
Nonylphenol	211	100	250	800	50	50	800
Tetrachlorobenzene	800	560	1000	4000	250	250	250

[1] LC50 = median lethal concentration

[2] NOEC = No Observed Effect Concentration

[3] LOEC = Lowest Observed Effect Concentration

2.4 Single Chemical Toxicity Identification Evaluations

The single chemical spiking experiments were designed to determine if the cause of toxicity could be characterized, then identified, in sediments containing a toxic concentration of one chemical. Several TIE treatments were applied for each chemical in order to compile lines of evidence indicating the cause of toxicity.

The single chemical TIE experiments were designed to proceed in two phases. First, five kilograms of formulated sediment was spiked with a single target concentration of the test compound. Spiked sediments were rolled for 24 hours at 15 °C, and then stored in darkness at 4 °C to allow equilibration. The sediment was then subjected to various solid-phase TIE treatments. If the solid-phase TIE correctly characterized the cause of toxicity, a subsample of sediment was analyzed to confirm the chemical concentration. During the preparation of the solid-phase treatments, interstitial water (IW) was extracted from the remainder of the untreated sediment and used to conduct an initial IW test. When the IW test produced significant toxicity (defined as <50% survival), sufficient additional sediment was spiked to produce interstitial water for all IW TIE treatments and synoptic chemical analyses. After equilibration, the IW was extracted and a TIE was conducted. Depending on the success of the TIE, an IW sample was analyzed to confirm the chemical concentration. In this approach, solid-phase TIEs and initial

IW toxicity tests were started on the same day. Because of the effort required to produce sufficient IW, these TIEs were conducted only after sufficient toxicity was confirmed in the initial test. Sediment for additional solid-phase and interstitial water TIEs was spiked and equilibrated as needed.

To confirm the efficiency of the TIE treatments at reducing or removing the toxicants, final sediment and interstitial water tests were performed in conjunction with chemical analyses. For solid-phase TIEs, the chemical concentrations in the sediment and the eluate of the solid-phase amendment were measured. Interstitial water concentrations in the baseline, the solid-phase extraction (SPE) column rinsate, and the SPE column eluate were measured.

2.5 Multiple Chemical Toxicity Identification Evaluations

Our primary goal in the multiple chemical experiments was to evaluate whether existing and modified sediment TIE techniques could identify specific toxicants in samples containing complex contaminant mixtures. The three mixture experiments each contained all five of the study chemicals, with two chemicals spiked at toxic concentrations and the remaining chemicals spiked at approximately one-quarter TU. The TIE treatments that were most effective in the single chemical experiments were used individually, sequentially, or in combination to remove toxicity from both sediment and interstitial water, and to identify toxic fractions in the eluates from resins and columns. Each sample was subjected to several treatments to generate multiple lines of evidence in each TIE.

These experiments proceeded in the same manner as the single chemical experiments. Approximately 15 kilograms of formulated sediment was spiked with all five test chemicals. Two of the chemicals were spiked at the original single chemical concentrations, while the other three chemicals were spiked at approximately one-quarter of a toxic unit. Spiked sediments were rolled for 24 hours at 15 °C, and then stored in darkness at 4 °C until use. Solid-phase and interstitial water TIEs were conducted simultaneously. Target treatments are listed in Figure 2-1, but the following treatments were used most often for solid-phase tests: 10% percent SIR-300 addition and elution, 10% Ambersorb addition and elution, 20% zeolite addition, and a combination of these treatments. Interstitial water TIEs primarily employed zeolite, air stripping, EDTA, cation column extraction and elution, HLB column extraction and elution, and combinations of these treatments.

Toxicological and chemical measurements were combined in the final solid-phase and interstitial water tests for each mixture. For solid-phase TIEs, chemical concentrations were measured in baseline sediments and amendment eluates. For interstitial water TIEs, concentrations were measured in the baseline samples, column rinsates, and column eluates. TIE results for solid-phase and interstitial water are each presented in two tables, one for the preliminary test and one for the evaluation of extraction efficiency.

2.6 Ambient Samples

After the sediment TIE methods were evaluated using spiking experiments, six toxic ambient samples were collected to validate the methods. Three marine and three freshwater sediments were collected from four sites recommended by the PSC, and from two additional sites in California.

Three marine samples were collected from southern California (U.S.). Collection methods followed ASTM (2000). Switzer Creek (SZ) is located in San Diego Bay, CA. The site is located between the north side of the 10th Avenue Marine Terminal and the Cambell Shipyard Piers at the mouth of Switzer Creek. This station is at the base of an urban storm drain receiving nonpoint source inputs. Sediment was originally collected for a San Diego Bay TMDL project on April 23, 2004 (Anderson et al., 2004). Preliminary Phase I TIEs from this study suggested toxicity was caused by organic contaminants. There was sufficient sediment remaining to continue further characterization and identification of the causes of toxicity. The second ambient marine sediment sample was collected from Consolidated Slip (CS), located at the base of the Dominguez Channel in Long Beach Harbor; CA. Dominguez Channel is an urban drainage that receives runoff from the Los Angeles basin. Sediments from this site have demonstrated consistent toxicity in a number of previous studies (Anderson et al., 2001; Moore et al., 2004; Young et al., 2005). Sediment was collected by Weston Solutions (Carlsbad, CA) in 2001 and obtained by the project team at the Marine Pollution Studies Laboratory (MPSL) in February 2005. The third ambient marine sediment sample was collected from Upper Newport Bay (UN) in southern California. This station receives input from San Diego Creek, which drains urban areas and rural rangeland, and receives some agricultural and greenhouse runoff. Sediment was collected on April 12, 2005.

An additional solid-phase TIE was conducted using the sediment from station SZ to evaluate loadings of contaminants into amphipod tissues during the exposure. SZ sediment was diluted to 10% strength and tested alone, with SIR-300, and with Ambersorb. Surviving amphipods were preserved at -20ºC and analyzed for pyrethroids, organochlorines, PCBs, and metals. Chemical reporting limits are listed in Tables C-21 to C-24. Because the expected response of *E. estuarius* in 10% SZ sediment was 50% survival, the replication was increased from five to 25 to increase the amount of tissue recovered for analysis. The minimum amount of tissue necessary for basic analyses was 1g wet weight, but obtaining additional tissue improves detection limits.

Two freshwater ambient samples were collected in California, and the third came from Indiana. San Diego Creek (SD) is approximately 0.5 km upstream of the Upper Newport Bay station described above. Sediment was collected at this station on April 12, 2005. Sediment from Indiana Harbor Canal (IH) was collected by Scott Cieniawski of the U.S. EPA in Chicago, IL on June 9, 2005. Indiana Harbor receives industrial effluent inputs and this site has been the subject of ongoing sediment quality studies (Ingersoll et al., 2002; MacDonald et al., 2002). The third ambient freshwater sediment came from Alisal Creek (AC), which is located at the southern edge of the City of Salinas, CA. The slough receives inputs from urban and agricultural nonpoint sources. In previous studies, this sediment has been consistently toxic to *Hyalella azteca*. This sample was collected on July 19, 2005. Upon arrival at the MPSL, all sediment was stored in the dark at 4 ºC until used. Approximately 40 L of surficial sediment were collected from each station and homogenized.

Initial toxicity tests were conducted on homogenized solid-phase samples and on interstitial water that was extracted via refrigerated centrifuge at 2500 *g*. Tests were conducted on a dilution series of both solid-phase sediment and interstitial water (described above) to determine the magnitude of toxicity. Dilution tests and preliminary TIEs allowed for definitive tests to be performed at the highest possible concentration. All solid-phase TIEs were conducted on a single concentration of sediment. For the marine ambient sample TIEs, Switzer Creek (SZ) sediment was diluted to 25%, Consolidated Slip (CS) to 1%, and Upper Newport Bay (UN) to

10%. For the freshwater TIEs, San Diego Creek (SD) and Alisal Creek (AC) were tested at 100% and Indiana Harbor (IH) was diluted to 25%. Interstitial water and solid-phase TIEs were conducted on all samples except for sediments from SZ and SD, which were only subjected to solid-phase TIEs, due to a lack of interstitial water toxicity. In the case of station SZ, the undiluted sediment was pre-sieved through a 280 μm screen prior to dilution to facilitate the separation of the TIE amendments (procedure described in Appendix A).

2.7 Specific Methods Considerations

Solid-phase and interstitial water TIE procedures are outlined in Appendix A, but specific methods considerations are discussed below that will aid in the interpretation of the TIE experiments.

In the solid-phase exposures, the SIR-300 and Ambersorb amendments are added to sediment to reduce the toxicity of metals and organic compounds, respectively. Upon test termination, these amendments are separated from the sediment for elution (method outlined in Appendix A). SIR-300 is eluted with acid and Ambersorb is eluted with acetone. The acid fraction is added to 100 mL of dilution water, neutralized with sodium hydroxide, and then used in a toxicity test as the SIR-300 eluate sample. The acetone fraction is evaporated to 1 mL, added to 100 mL of dilution water, and then used in a toxicity test as the Ambersorb eluate sample. Unlike the interstitial water SPE columns, in which the acid or solvent fraction is reconstituted at the same volume as the amount of sample passed through the column, the volume of water that these fractions are added to does not directly relate to the sediment or interstitial concentrations of contaminants, or to the amount of sediment in which the amendment was originally placed. Also, the amendment is not assumed to have 100% adsorbant efficiency. Adjusting the final concentration of water that the fraction is added to may increase or decrease the toxicity of the eluate. Therefore, a standard volume of 100 mL was chosen because that volume was the minimum amount necessary to conduct the toxicity exposure and the amphipods can safely tolerate a concentration of 1% solvent. Adding the solvent to a minimal amount of water also maximizes the toxicity and chemistry signals. The presence of toxicity and contaminants provide lines of evidence for the original cause of toxicity in the sediment. It is also not assumed that the resin was fully recovered from the sediment, or that the rinsing process compromised the resin.

Formulated sediment was used in all spiking experiments. Solid-phase TIE treatments were evaluated by adding an amendment directly to the spiked sediment. To account for dilution, a separate dilution blank was prepared that contained the same amount of spiked formulated sediment diluted to the appropriate concentration with clean formulated sediment. When necessary, freshwater ambient samples were diluted with formulated sediment, while marine ambient samples were diluted with sediment from the amphipod collection site. Similarly, for the ambient TIEs freshwater dilution blanks were created with formulated sediment and marine dilution blanks were created with amphipod collection site sediment. Use of the term "control" depends on the type of test conducted.

In the early interstitial water TIEs, filtration was used to remove particles from the samples (U.S. EPA, 1991). This proved to be an effective treatment for removing organic contaminants as well as particles. Because interstitial water was obtained through centrifugation, few large particles remained in the samples. As the study progressed, additional centrifugation was used instead of filtration to remove suspended material, and as a pre-treatment step for the column treatments. Ankley et al., (1991) suggest using centrifugation at 20,000-30,000 *g* instead filtration for pre-column treatment, but the samples in this study were centrifuged at 3200 *g*.

Similarly, early interstitial water TIEs used the C18 column for the solid phase extraction of non-polar organics (Varian Bond Elut®, 6 mL, 500 mg, Varian Inc., Palo Alto, CA, U.S.). Later experiments utilized an HLB column (Oasis Hydrophilic-Lipophilic Balance®, 6 mL, 500 mg, Waters Corporation, Milford, MA, U.S.). The hydrophilic and lipophilic Oasis column removes many polar contaminants as well as the non-polar contaminants that are sorbed by the C18 column (Waters, 2001). The HLB column has other advantages that include increased surface area, lack of pH sensitivity in extraction efficiency, and the ability to dry out between manipulations, if necessary. The generic method of loading and eluting the HLB column recommended by the manufacturer was the most successful (see Appendix A). Other researchers have observed increased toxicity in some HLB column eluates, indicating the HLB packing materials could be contributing to toxicity with some interstitial water samples, and are planning collaborative studies with the manufacturer to improve organic extractions for TIEs (SAIC, 2003).

Graduated pH treatments were used on interstitial water samples to characterize toxicity caused by pH-dependent toxicants such as ammonia, sulfide, and some metals. Graduated pH treatments provided useful data in single chemical interstitial water TIEs with ammonia in both freshwater and marine TIEs and with copper in the marine TIE. Although blank survival was acceptable in the spiked sediment TIEs with the organic contaminants, complete mortality was observed in most of the dilutions. The cause of the mortality in the dilutions is unknown. Because of the excessive mortality related to the organic contaminants, graduated pH treatments were not used in the multiple chemical TIEs or most of the ambient TIEs. Graduated pH data from the organic contaminant TIEs will not be discussed further, but the data are provided in Appendix B.

The addition of the green alga *Ulva lactuca* to sediment overlying water or interstitial water was not evaluated. Burgess et al. (2003) reported that zeolite addition and *Ulva* addition produced similar amphipod survival in ammonia-spiked sediments. It was also determined that commercially purchased zeolite would be easier to obtain for most testing laboratories likely to use these methods.

2.8 Data Analysis and Interpretation

Phase I TIEs are designed to broadly characterize the cause of toxicity as a metal, an organic, ammonia, or a combination. Whether interpreting data from spiked sediment or from an ambient sample, the first step is to qualitatively determine which treatments reduced or increased toxicity. Solid-phase TIE data were interpreted by first evaluating treatment blanks to determine if sample manipulations added toxic artifacts. The organism response in any given treatment was then qualitatively compared to the organism response in the baseline. Statistical comparisons between treatment and baseline were made using a separate variance t-test (alpha = 0.05). In interstitial water TIEs, treatment blanks also were evaluated first to determine if sample manipulations added toxic artifacts. Mortality in the various TIE treatments was then assessed at different time intervals to evaluate their effectiveness during the 10-day exposure. Treatment data were then compared to one another using the toxic unit approach. Toxic units (TU) were calculated by dividing 100 by the LC50 calculated from each treatment dilution series. A lower

TU value indicates a treatment has been effective in reducing toxicity. Although IW tests were observed and counted daily, as a general rule LC50s and TUs were calculated for Days 4, 7, and 10, but occasionally LC50 and TUs were calculated for other days. This allowed an examination of different stages of the IW TIEs and helped to determine when the best resolution among treatments occurred. The data that provided the best resolution are presented in the results below.

Treatments used in single chemical TIEs are evaluated at the end of each result section. The solid-phase and interstitial water treatments that did and did not reduce toxicity are listed, as well as a discussion on which treatments might be improved. For the multiple chemical TIEs, an evaluation was completed on whether or not the causes of toxicity could be differentiated between the two dominant chemicals. If a treatment successfully reduced toxicity, or returned toxicity with a significant portion of a dominant chemical, it was considered a line of evidence contributing to the overall cause of toxicity. The success of the ambient TIEs was measured based on similar lines of evidence. Solid-phase and interstitial water TIE results were combined with chemical analyses of the sediment, interstitial water, and treatments that returned toxicity to create a weight of evidence for characterizing and identifying the cause of toxicity.

CHAPTER 3.0

RESULTS AND DISCUSSION

3.1 Single Chemical Toxicity Identification Evaluations

In the following sections, results are presented for each single-chemical TIE, followed by a summary discussion of the observed effectiveness of TIE treatments for those toxicants. TIE results are organized from each chemical into two or three tables. Results from the single chemical spiked solid-phase TIEs are presented in the first table. The interstitial water TIE results are presented in the second and, if necessary, third tables that included chemical confirmation. All applicable tables are provided in Appendix B.

3.1.1 Marine Sediment and Interstitial Water TIEs

Ammonia

The TIE evaluations for ammonia used *E. estuarius* with 96-hour water-only LC50s of 278 mg/L total ammonia and 1.35 mg/L unionized ammonia (MPSL unpublished data). *E. estuarius* inhabits both the interstitial water and the overlying water, so measurements of IW ammonia would only provide a worse-case scenario for ammonia exposure in solid-phase sediment (Phillips et al., 1997). Unionized ammonia was calculated using the total ammonia concentration and an equation that takes into consideration pH, salinity and temperature (Phillips et al., 1997). The acceptability criteria for the *E. estuarius* protocol (U.S. EPA 1994c) sets an application limit of 0.8 mg/L for unionized ammonia. If this value is exceeded, the protocol states that ammonia may be a factor influencing toxicity, and that the cause of toxicity cannot be unambiguously ascribed to another toxicant. The 0.8 mg/L application limit is below the LC50 of 1.35 mg/L derived at MPSL. In order to interpret the results of the ammonia removal treatments in the solid-phase experiments with *E. estuarius*, the unionized ammonia concentration in the overlying water was compared to the U.S. EPA application limit of 0.8 mg/L. For the IW tests, the interstitial water ammonia concentration was compared to the water-only LC50 of 1.35 mg/L.

Solid Phase

Survival of *E. estuarius* in ammonia-spiked sediment was 20%, and the addition of zeolite increased survival to 56% (Table B-1). Survival in control sediment with zeolite was not significantly different from survival in the control, and survival in spiked sediment mixed with control sediment (20% dilution control to account for addition of the amendment) was not significantly different from the baseline test. Addition of zeolite reduced the total ammonia concentration in the overlying water from 138 mg/L to 36.1 mg/L, and the unionized ammonia from 2.03 mg/L to 0.828 mg/L. The concentration of unionized ammonia in the zeolite treatment was therefore reduced to approximately the recommended overlying water threshold for *E. estuarius* in 10 d exposures (0.8 mg/L, U.S. EPA, 1994). This treatment was effective in reducing ammonia toxicity in the context of a solid-phase sediment TIE.

Interstitial Water

The baseline test of the ammonia IW TIE contained 690 mg/L total ammonia and 1.4 TU (Table B-2). The concentration of unionized ammonia in the baseline test was 1.944 mg/L, about 1.4 times the LC50 for *E. estuarius* (1.35 mg/L, MPSL unpublished data). Passing the IW through a zeolite column reduced the total ammonia to 10 mg/L, the unionized ammonia to 0.07 mg/L, and the toxicity to less than 1 TU. Although air stripping also reduced the toxicity to less than 1 TU, the total ammonia concentration was only reduced to 336 mg/L in this treatment and the unionized ammonia concentration was 4.69 mg/L, well above the LC50 (Table B-2). The reduction of toxicity in the presence of high unionized ammonia toxicity might be an artifact resulting from the pH adjustment process used in the air stripping treatment. Differences in pH values could account for the lack of toxicity because as pH increases the amount of unionized ammonia increases, but the toxicity of the unionized ammonia decreases (U.S. EPA 1993). There might have been a temporary disequilibrium that accounted for the lack of toxicity. The graduated pH treatments were successful in characterizing ammonia toxicity by demonstrating increasing TUs and increasing unionized ammonia with increasing pH.

The IW treatments were more successful at characterizing and identifying ammonia toxicity than the solid-phase treatments. The zeolite column removed the majority of ammonia and toxicity, while a high ammonia concentration was present after air stripping at pH 10. After adjusting the air stripping dilutions to ambient pH, there was still a high concentration of unionized ammonia, but low observed toxicity. The graduated pH treatments supported the weight-of-evidence. Burgess et al. (2004) evaluated zeolite as an ammonia removal tool in marine waters, and while they observed similar removal, they also observed complete removal of copper that had been spiked in combination with ammonia (discussed further in mixture results).

Marine TIE Method Evaluation - Ammonia

Several treatments were successful at characterizing and identifying ammonia as the cause of toxicity in ammonia-spiked sediment experiments with *E. estuarius*. Zeolite was successful at reducing ammonia toxicity in both solid-phase and IW samples. Because this treatment potentially reduces toxicity caused by metals, it is recommended for use in concert with graduated pH treatments to provide two lines of evidence. Air stripping also successfully reduced toxicity and ammonia, and could be used in cases where metals might be contributing to toxicity. Graduated pH provided an additional line of evidence for the cause of toxicity.

Copper

Solid Phase

Four solid-phase TIEs were conducted using copper-spiked sediment (Table B-3). The first experiment used sediment spiked at 1000 mg/kg and failed to elicit a toxic response with *E. estuarius*. These data are not shown. Subsequent tests used sediment spiked at 2000 mg/kg. In Test 2, there was 8% survival in the baseline test. Addition of 10% and 20% SIR-300 increased the survival to 88% (Table B-3). The control with 20% SIR-300 was not significantly different from the test control, and the dilution control demonstrated a similar response to the baseline. In Test 3 the SIR-300 resin was eluted with acid and the eluate was added back to clean dilution water. Complete mortality was observed in the baseline test, and the addition of 10% SIR-300 increased survival to 84%. In this experiment the dilution control also demonstrated a significant increase in survival. Two attempts at eluting copper from the SIR-300 resin failed to return toxicity. In test 4, baseline survival was not as low as the previous two experiments (56%), but survival was significantly increased by the addition of SIR-300. The resin addition also caused a dilution effect in this experiment. Elution of the SIR-300 again failed to return toxicity.

The concentration of spiked copper was measured in sediment from Tests 3 and 4. Sediments spiked with 2000 mg/kg total copper contained 2035 mg/kg in Test 3 and 1610 mg/kg in Test 4 (Table B-3). Although SIR-300 elution add back to dilution water resulted in an elution test concentration of 112 mg/L copper, and this concentration was greater than the water only LC50 of 33.3 mg/L, no significant toxicity was observed in the eluate sample. Subsequent experiments with SIR-300 elution demonstrated that high concentrations of copper could be eluted, but toxicity was not observed until elution test concentration reached 174 mg/L copper. It appears that copper eluted from SIR-300 with hydrochloric acid is not bioavailable when spiked into seawater, for reasons currently unknown. This could be clarified by measuring cupric ion activity in the spiked seawater, but this was beyond the scope of the current study.

In two of the three solid-phase tests, the dilution control results were not significantly different from the SIR-300 treatment results, suggesting the reduction of toxicity might have been caused by dilution from the addition of the resin. However, significant dilution effects were not observed in Test 2, indicating that in this test, at least, the reduction in toxicity was due to copper binding to the SIR-300 resin rather than dilution of the baseline sediment.

Interstitial Water

The IW TIE was conducted with several treatments designed to reduce metal toxicity (Table B-4). The baseline sample contained 2.1 TU. Toxicity was reduced by addition of both EDTA and sodium thiosulfate, and by the cation column. No treatment completely removed toxicity. There was no toxicity in the cation column eluate when reconstituted in clean dilution water. The graduated pH treatment demonstrated decreasing toxicity with increasing pH. In the second IW TIE, the baseline contained 8 TU and 64.4 mg/L copper (Table B-5). The cation column reduced the toxicity to 1.4 TU and the copper concentration to 43.6 mg/L, which is 1.3 times the *E. estuarius* LC50 (33.3 mg/L; MPSL unpublished data). The cation column did not completely remove toxicity, but when the cation eluate was added back to clean dilution water, the resulting test solution had 2.4 TU and 34.7 mg/L copper.

EDTA and the cation column successfully reduced copper toxicity to *E. estuarius*, but did not completely remove it (Table B-4). EDTA doses of 60 and 120 mg/L were used, enough to chelate approximately 22-44 mg/L copper. The concentration of copper in the interstitial water was 51.7 mg/L, so remaining toxicity was expected. Higher concentrations of EDTA could be used, but the EDTA tolerance threshold of *E. estuarius* is unknown. The cation column was not expected to remove all of the copper from the interstitial water because the concentration of copper in the samples was approximately 300 times the concentration used by Burgess et al., (1997). The cation column eluate successfully returned copper to clean dilution water, but because of the high tolerance of *E. estuarius* to copper, it was difficult to add back sufficient copper to cause toxicity. The graduated pH treatment appears to provide another line of evidence that copper was causing toxicity, but these results were inconsistent with previous reports. Lower survival was observed at lower pH in the tests. Ho et al. (1999) also reported that copper toxicity to the amphipod *Ampelisca abdita* decreased with decreasing pH, while survival of two other species increased when pH was lowered.

The copper concentration in the first IW TIE was 51.7 mg/L, approximately 1.5x the nominal 10-day water-only LC50 (33.3 mg/L, MPSL unpublished data). The concentration in the second IW TIE was 64.4 mg/L (Table B-5). Because the cation column only reduced this concentration to 43.6 mg/L, residual toxicity was observed in the column rinsate. The cation column eluate returned 34.7 mg/L, and survival was 40% in the 100% concentration using this treatment. Elution of the cation column with hydrochloric acid did not seem to reduce copper bioavailability as observed with elution of the SIR-300. However, the copper mass removed by the column from the baseline IW was not equivalent to the copper mass after cation column elution.

Marine TIE Method Evaluation - Copper

A combination of solid-phase and IW TIEs were able to characterize and identify copper toxicity. SIR-300 successfully reduced copper toxicity, but the resin elution failed to return toxicity even though more than 3 TUs of copper were added back. Further research is needed to determine why the elution of SIR-300 with hydrochloric acid reduced the bioavailability of copper. EDTA, sodium thiosulfate, and the cation column reduced IW toxicity, and the cation column eluate returned toxicity in the second IW TIE. The gradated pH results provided some evidence of copper toxicity, but the organism response to metal toxicity with graduated pH is species specific, and might not provide a strong line of evidence. Further research is needed to refine the cation column technique to avoid breakthrough of copper.

Fluoranthene

Solid Phase

Five solid-phase TIEs were conducted with fluoranthene-spiked sediment. In the first two TIEs, the fluoranthene spiking concentration was based on an LC50 derived from the initial formulated sediment that contained 0.39% TOC. All subsequent experiments used formulated sediment containing 0.75% TOC, but the higher TOC reduced fluoranthene bioavailability, so sediments had to be spiked with higher fluoranthene concentrations. Data are therefore not presented from the first two TIEs that used the lower fluoranthene concentration.

Complete mortality was observed in the baseline sediment in Test 3 (Table B-6). Addition of 10% Ambersorb and 10% PCC each increased survival to 88%. Toxicity was reduced to a lesser extent by comparable sample dilution with control sediment. No toxicity was observed in the Ambersorb eluate. A toxic concentration of fluoranthene could not be eluted from the Ambersorb in Test 4. Again, Ambersorb and PCC significantly increased survival over the baseline, but there was notable mortality in the control (64% survival) attributed to low survival in one replicate (data not shown). The acetone elution of the Ambersorb successfully returned toxicity to clean dilution water. In the fifth solid-phase TIE, the efficiency of Ambersorb elution was assessed with acetone by measuring fluoranthene in the eluate. As with Test 4, Ambersorb significantly increased survival in the Test 5 sample, and the elution successfully returned toxicity.

The concentration of fluoranthene in the Test 4 baseline sample was 50.0 mg/kg, approximately half the nominal concentration (Table B-6). The concentration in the Test 5 baseline sample was 20.6 mg/kg, about a quarter of the nominal concentration. Although the measured concentrations were below the nominal concentrations, significant toxicity was observed in both baseline exposures. Sufficient fluoranthene was eluted from the Ambersorb in Tests 3 and 4 to return significant toxicity to dilution water. In Test 5, the final concentration of the fluoranthene in the eluate was 241 µg/L (Table B-6). Water-only LC50s were not calculated as part of this study, but Boese et al. (1997) report the LC50 for fluoranthene is greater than 70 μg/L.

Ambersorb and PCC reduced fluoranthene toxicity with some dilution effect. Ho et al. (2004) reported that the addition of PCC removed all toxicity from endosulfan-spiked sediments, and from PCB and PAH-contaminated field sediments. Previous studies with Ambersorb have shown that it reduces toxicity caused by PAHs (Kosian et al., 1999; West et al., 2001). Because of its size, Ambersorb can be recovered from sediment via sieving. This makes it possible to elute the resin with solvent, an important step in the Phase II TIE process. Elution is not possible with PCC. Although the procedures require refinement (discussed below), Ambersorb recapture and elution present an important advancement in solid-phase TIEs, and provide another line of evidence for the cause of toxicity.

Interstitial Water

The baseline test in the IW TIE contained 3.6 TU (Table B-7). Filtration reduced toxicity to 1.2 TU, and aeration slightly reduced toxicity. Because filtration was used as a pre-treatment step prior to solid-phase extraction with the C18 and HLB columns, both columns further reduced toxicity to <1 TU. The C18 eluate did not return toxicity, but the HLB eluate was significantly toxic (8 TU). Although significant toxicity was observed in this fluoranthene-spiked sample, fluoranthene was not detected by GC-MS analysis. Quality assurance measurements for recovery of fluoranthene were acceptable, so the reason for the lack of PAH detection in the IW is unclear.

The second IW TIE was designed to test the solvent extraction efficiency of the HLB column (Table B-8). Significant toxicity was observed in the baseline sample (1.6 TU), and the column reduced toxicity to <1 TU, but no toxicity was observed in the HLB eluate. The IW contained 73.2 μg/L of fluoranthene, and less than 4 μg/L was measured in the column rinsate. The column eluate contained only 1.4 μg/L, indicating inadequate elution of fluoranthene from the column using methanol.

Results of the filtration treatment confounded the interpretation of the first interstitial water TIE. Filtration is effective at removing particles from the sample, but because IW is extracted via centrifugation, minimal particles are present. Filtration was therefore omitted in later TIEs due to concerns of sorption of chemicals to the filter. Both column treatments reduced toxicity to <1TU, but toxic concentrations of fluoranthene were only eluted from the HLB column, not the C18 column. The second HLB elution was not successful, even though the column had removed almost all of the fluoranthene from the sample. The generic procedure for column elution recommended by the manufacturer (Waters, Inc., Milford, MA) was followed. Waters Inc. also recommend elution with methylene chloride for more hydrophobic compounds. Further research is needed to determine if this procedure is amenable to TIEs with *E. estuarius*. Regardless of this missing Phase II step, when combined with synoptic chemistry, the HLB column provided a useful line of evidence for identifying fluoranthene toxicity in the interstitial water.

Marine TIE Method Evaluation - Fluoranthene

Solid phase and IW TIEs were able to characterize and identify fluoranthene toxicity. Ambersorb and PCC successfully reduced fluoranthene toxicity, and the acetone eluate of Ambersorb returned toxicity and fluoranthene to dilution water. The HLB column reduced toxicity in both IW tests, but only returned toxicity in one test. The HLB column loading and elution methods need to be refined so that organic chemicals can be consistently loaded without breakthrough, and eluted for toxicity testing and chemical analysis.

Nonylphenol

Solid Phase

Baseline survival in the first solid-phase TIE with nonylphenol was 0% (Table B-9). Addition of 5% Ambersorb increased survival slightly, while addition of 20% PCC increased survival to 96%. Treatment blanks had acceptable survival, and sample dilution with control sediment affected toxicity only slightly. In Test 2, additional masses of Ambersorb and PCC were tested. Additions of both Ambersorb and PCC significantly increased survival in nonylphenol-spiked sediments. Ambersorb was eluted with acetone, and the eluate tested in the third solid-phase TIE. The 5% Ambersorb addition increased survival to 48%, and significant toxicity was returned to clean dilution water when the isolated Ambersorb was eluted with acetone. Test 4 used a 10% Ambersorb addition, but only increased the survival from 0% to 20%. The Ambersorb eluate in this test caused complete mortality when it was spiked into clean dilution water.

The baseline concentrations of nonylphenol measured in Tests 2 and 3 were 99 and 108 mg/kg, respectively. These were less than half the nominal spiked concentrations (Table B-9). The concentration of nonylphenol in Test 4 sediment was 215 mg/kg, and the concentration in the Ambersorb eluate from this test was 33 mg/L, well above the LC50 of 227 μg/L (Hecht and Boese, 2002).

As observed in TIEs with fluoranthene, Ambersorb was less effective at reducing nonylphenol toxicity than PCC. However, Ambersorb provides the critical ability to recover toxicants and toxicity through elution, providing an important line of evidence for the nonylphenol TIE.

Interstitial Water

In the first TIE with IW extracted from nonylphenol-spiked sediment, the baseline contained 1.4 TU (Table B-10). Filtration reduced toxicity to 1.1 TU, and the C18 column reduced to <1 TU. Elution of the C18 did not return toxicity to clean dilution water. The HLB column was not evaluated in this TIE. The second IW TIE was designed to test the extraction efficiency of the HLB column. The baseline sample contained 4.8 TU, and the HLB column reduced the toxicity to 1.9 TU (Table B-11). The HLB eluate successfully returned 2.8 TU to clean dilution water. There were 350 μg/L nonylphenol in the baseline. The HLB column reduced the nonylphenol concentration to 10 μg /L, and the HLB eluate contained 292 μg/L.

Marine TIE Method Evaluation - Nonylphenol

Based on these results, the solid-phase and IW TIEs were capable of identifying nonylphenol. Ambersorb and PCC reduced solid-phase toxicity and acetone elution of the Ambersorb returned toxicity and nonylphenol to dilution water. The HLB column performed better than the C18 column by reducing and returning toxicity and nonylphenol to dilution water. As with the fluoranthene TIEs, the HLB column loading and elution methods need to be refined so that organic chemicals can be consistently loaded without breakthrough, and eluted for toxicity testing and chemical analysis.

Tetrachlorobenzene

Solid Phase

Baseline survival in the first tetrachlorobenzene TIE was 0% (Table B-12). Addition of 5% Ambersorb increased survival to 88%. There was acceptable survival in the Ambersorb control, and sample dilution with control sediment affected toxicity only slightly. Additional concentrations of Ambersorb and PCC were used in Test 2. Five percent Ambersorb and 10% PCC did not increase survival as well as 10% Ambersorb and 20% PCC (Table B-12). Elution of the 10% Ambersorb recovered negligible toxicity, but this procedure was attempted early in the method development process, and subsequent tests proved to be more successful. Test 2 treatment blanks and dilution blanks were acceptable. In Test 3, 10% Ambersorb increased survival from 4% to 88%. Test 3 elution of the Ambersorb with acetone successfully added back toxicity to clean dilution water. This eluate contained 6.9 mg/L tetrachlorobenzene.

The concentration of tetrachlorobenzene in Test 2 sediment was 402.1 mg/kg, approximately half the nominal spiked concentration of 800 mg/kg. The Test 3 concentration of tetrachlorobenzene was only 60.5 mg/L, less than 10% of the nominal spiked concentration. Despite these apparently low concentrations, significant mortality was observed in all baseline sediment exposures.

Interstitial Water

The baseline test in the first tetrachlorobenzene IW TIE contained 1.4 TU (Table B-13). Filtration and aeration completely removed the toxicity. Because filtration was used as a pre-treatment step for the C18 column, the column did not further reduce toxicity and there was no tetrachlorobenzene returned in the column eluate. Baseline toxicity in the second IW TIE accounted for 2.8 TU and the IW contained 1722 μg/L of tetrachlorobenzene (Table B-14). The HLB column reduced the toxicity to 1.6 TU and the IW concentration of tetrachlorobenzene to 6.9 μg/L. Although there was less than 1 TU, there was 40% mortality in the 100% treatment when the HLB eluate was returned to clean dilution water. The eluate sample contained 137.4 μg/L tetrachlorobenzene.

The previous experiments demonstrated the utility of the HLB column in extraction of IW organics and solvent elution for Phase II TIE treatments. In the second IW TIE with tetrachlorobenzene, the HLB column successfully removed most of the chemical, but not all of the toxicity (Table B-14). The remaining toxicity in the column rinsate confounds the interpretation of the data since it contained only 6.9 μg/L tetrachlorobenzene. Further research is needed to determine if the HLB column is an appropriate SPE tool for removing tetrachlorobenzene from marine IW. This should include examination of the extraction efficiency of the HLB column combined with measures of breakthrough.

Marine TIE Method Evaluation - Tetrachlorobenzene

Using a combination of both Ambersorb and PCC amendments, along with Ambersorb elution and chemistry, the cause of toxicity was successfully characterized and identified as tetrachlorobenzene. Ambersorb and PCC successfully reduced tetrachlorobenzene toxicity, and the Ambersorb elution in the third test returned a significant amount of toxicity and tetrachlorobenzene to dilution water. The HLB column removed toxicity and tetrachlorobenzene, but further refinement is necessary to improve the elution process.

3.1.2 Freshwater Sediment and Interstitial Water TIEs

Ammonia

Previous testing of interstitial water extracted from ammonia-spiked sediments provided *H. azteca* 96-hour water-only LC50s of 160 mg/L for total ammonia and 5.44 mg/L for unionized ammonia (MPSL unpublished data). The total ammonia value falls within the range reported by Ankley et al. (1995) and Besser et al. (1998), but the unionized value was approximately three times higher than the LC50 reported by Besser et al. (1998). *H. azteca* are epibenthic organisms that inhabit both the IW and overlying water in sediment toxicity tests. There is no guidance for ammonia tolerance in sediment exposures in the test protocol (U.S. EPA 2000), but Ankley et al. (1995) is cited for water-only LC50 values over a range of pH values. The LC50 of 5.44 mg/L was used to interpret interstitial, overlying, and treatment ammonia concentrations in all TIEs because it was derived from tests conducted with MPSL dilution water.

Solid Phase

Two solid-phase TIEs were conducted with *H. azteca* in ammonia-spiked freshwater sediment. In both cases the addition of 20% zeolite to the sediment significantly increased survival (Table B-15). The total ammonia concentrations in the overlying water differed markedly between the two experiments (307 mg/L vs. 90.2 mg/L). The addition of zeolite reduced the overlying water total ammonia and unionized ammonia by an average of 40%. The unionized ammonia concentrations were reduced below the LC50 of 5.44 mg/L in both experiments (Table B-15). No toxicity was observed in the zeolite control and minimal dilution effects. Average dilution of ammonia with 20% zeolite addition was 9%. These results agree with those of Besser et al. (1998), who found greater than 70% ammonia reduction in the interstitial waters of spiked sediments after zeolite addition. The interstitial water concentrations in these experiments were not measured, but there was a significant toxicity reduction associated with the decrease in overlying water ammonia concentration.

Interstitial Water

The IW contained 1096 mg/L total ammonia, 12.8 mg/L unionized ammonia, and 5.7 TU (Table B-16). Extraction of the sample with a zeolite column significantly reduced the toxicity to 1.6 TU and the total ammonia to 567 mg/L, but the unionized ammonia concentration (14.3 mg/L) remained well above the LC50 (5.44 mg/L). Unionized ammonia concentration is a function of pH, temperature and salinity. The pH of this treatment was 7.72, driving a higher fraction of ammonia to the unionized form, but as more unionized ammonia is created, the less toxic it becomes (U.S. EPA 1993). Although the total ammonia was reduced by approximately 50%, the unionized ammonia remained elevated. The air stripping treatment reduced toxicity to less than 1 TU and the total ammonia to 290 mg/L, but the unionized ammonia concentration also remained above the LC50. As was observed in exposures with *E. estuarius*, the reduction of toxicity without the reduction of unionized ammonia might be an artifact of the pH adjustment process in the air stripping treatment. Adjustment of the interstitial water to pH 10 and back to ambient pH might have caused a temporary lack of equilibrium that caused a lack of toxicity. As expected, toxicity increased with increasing pH and unionized ammonia concentrations in the graduated pH treatments. Toxicity ranged from 7.9 TU at pH 7 to 20 TU at pH 9. Total ammonia concentrations did not increase, but unionized ammonia increased from 5.4 mg/L at pH 7 to 303 mg/L at pH 9 (Table B-16).

Freshwater TIE Method Evaluation - Ammonia

Zeolite addition is an effective treatment for reducing ammonia concentrations and toxicity in solid-phase exposures. When used with other solid phase treatments, such as Ambersorb or SIR-300, this procedure provides useful data on the contribution of ammonia to sediment toxicity. Zeolite addition successfully reduced toxicity in both solid-phase and IW tests with *H. azteca*, but air stripping was more successful in the IW matrix. It is not known why there are elevated concentrations of unionized ammonia without an increase in toxicity in the air stripping treatments. The graduated pH treatment also provided a line of evidence that ammonia was contributing to toxicity. In IW samples where ammonia is the primary contaminant, a combination of these TIE treatments should be adequate for characterizing and identifying ammonia as the cause of toxicity.

Copper

Solid Phase

Four solid-phase TIEs were conducted with *H. azteca* exposed to copper-spiked freshwater sediment (Table B-17). In Test 1, addition of 20% SIR-300 increased survival from 0% to 56%. There were negligible dilution effects from the resin addition, but significant toxicity was observed in the SIR-300 blank, indicating the resin negatively affected the organisms. All water quality parameters in this experiment, including pH, were within acceptable ranges. In Test 3, addition of 10% SIR-300 increased survival in copper spiked sediment to 62% with no blank toxicity. Toxicity was recovered in the acid eluate of the 10% SIR-300 addition; with 0% survival in the eluate treatment. Similar results were observed in Test 4. All treatment blanks and controls had acceptable survival except for the second SIR-300 elution control in Test 4 (13% survival).

Sediments were spiked with 1000 mg/kg copper in these TIEs. Measured concentrations were 1065 mg/kg copper in Test 3 and 753 mg/kg copper in Test 4 (Table B-17). The SIR-300 elution in Test 4 added back 50.2 mg/L copper, which was an order of magnitude higher than the LC50, and there was 13% survival in this treatment. When the SIR-300 elution treatment was used with *E. estuaries*, toxicity was rarely observed. This could have been due to reduced copper bioavailability associated with spiking into seawater. Reduced copper bioavailability might also explain the observation of higher than expected survival in the resin eluate experiments with *H. azteca*. Although elution of SIR-300 with hydrochloric acid might affect the bioavailability of copper, the resin successfully removed and returned copper and toxicity, and should be considered an effective tool in determining the cause of toxicity for sediments containing elevated concentrations of divalent cations.

Interstitial Water

Several treatments in the interstitial water TIE reduced copper toxicity (Table B-18). Both EDTA treatments reduced toxicity to <1 TU. Toxicity in the sodium thiosulfate blanks confounded the ability to determine its effectiveness, and these treatments were not repeated. Both concentrations of EDTA (60 and 120 mg/L) successfully removed copper toxicity in the freshwater tests, whereas the 60 mg/L addition did not remove copper toxicity in the marine tests. The concentration of copper in this sample was 12.7 mg/L, approximately one quarter that used in the *E. estuarius* exposure. The 60 mg/L EDTA addition is designed to chelate up to 22 mg/L copper. The cation column completely removed toxicity and the eluate returned a significant amount of toxicity to clean dilution water. Graduated pH treatments were significantly toxic in every dilution and did not provide a line of evidence for copper toxicity.

The second IW TIE was designed to test the copper extraction efficiency of the cation column. The baseline test contained 7.4 TU and 14.2 mg/L copper (Table B-19). The cation column reduced the toxicity to <1 TU and the copper concentration to 1.1 mg/L. The cation eluate returned all of the toxicity 7.8 mg/L copper.

Freshwater TIE Method Evaluation - Copper

Addition and elution of SIR-300 characterized and identified copper toxicity in the solid-phase TIE. In the IW TIE, EDTA provided a useful line of evidence in the Phase I characterization of metals toxicity. The cation column, however, allowed both characterization and identification of copper as the toxicant. Additional research is required to improve the elution efficiency of the cation column for the Phase II TIE process.

Fluoranthene

Solid Phase

Five TIEs were conducted using *H. azteca* exposed to fluoranthene-spiked freshwater sediment. As with the marine fluoranthene evaluation, data from the first two tests are not presented because spiking levels were too low for the increased TOC content of the formulated sediment. Sediment for Test 3 was spiked with 120 mg/kg and resulted in 38% survival (Table B-20). Addition of 10 and 15% Ambersorb significantly increased survival, but the elution of the resin did not return toxicity. Addition of PCC had mixed results. Fifteen percent PCC addition reduced toxicity, but toxicity was observed in the PCC blank. The dose of 200 mg/kg fluoranthene in Test 4 resulted in 100% baseline sample mortality. Addition of 10 and 15% Ambersorb significantly increased survival, and toxicity was observed when Ambersorb eluate from each treatment was added back. The addition of 10 and 15% PCC partially reduced toxicity.

Fluoranthene in baseline samples was measured at 220 mg/kg in Test 4 and 48.7 mg/kg in Test 5 (Table B-20). Although the measured concentration of the sediment from Test 5 was lower, only 8% of the organisms survived. The concentration of fluoranthene eluted from the Test 5 Ambersorb was 530 µg/L, well above the water-only LC50 for *H. azteca* (44.9 µg/L; Suedel et al., 1993).

Ambersorb reduced fluoranthene toxicity to *H. azteca* to a greater extent than PCC. Kosian et al. (1999) and West et al. (2001) reported success at reducing the bioavailability of PAHs in sediment amended with Ambersorb 1500. Ambersorb 563 was used in the current study because Ambersorb 1500 is no longer produced. During this project it was discovered that Ambersorb 563 has also been removed from production. Currently, experiments are being conducted at MPSL to determine the efficacy of an alternative organic removal resin, Amberlite XAD-4 (Rohm and Haas, Spring House, PA, U.S.).

Interstitial Water

The baseline test in the first IW TIE contained 3.2 TU (Table B-21). Filtration reduced the baseline toxicity to 2.2 TU and aeration reduced it to 1.2 TU. The HLB column reduced the toxicity in this TIE to 2 TU, while elution of the column added back 1.3 TU. The C18 column removed slightly more toxicity (1.9 TU), but the C18 eluate was not toxic. The concentration of fluoranthene in the baseline test was 144.3 µg/L, much greater than the water-only LC50 of 44 μg/L (Suedel et al., 1993), but only moderate toxicity was observed in this sample (Table B-22). The IW used in this test was extracted from the sediment used in the fifth solid-phase test, and measured fluoranthene was approximately one-quarter of the nominal dose. The HLB column reduced the concentration of fluoranthene to 1.9 μg/L, but the elution of the column only recovered 2.1 μg/L, and was not toxic.

Freshwater TIE Method Evaluation - Fluoranthene

Addition of Ambersorb and PCC to fluoranthene-spiked sediment, and the acetone elution of the Ambersorb, successfully characterized and identified fluoranthene toxicity. The IW TIEs were less conclusive. The SPE columns reduced toxicity, but only one of the HLB columns returned toxicity. Further research is needed to improve the loading and elution of the HLB column.

Nonylphenol

Solid Phase

Five freshwater solid-phase TIEs were conducted with nonylphenol-spiked sediment (Table B-23). In the first test, toxicity was removed with 10 and 15% PCC, but only partial removal occurred with 5% Ambersorb addition. There was also a significant reduction of toxicity in the 15% dilution control. Results of Tests 2 and 3 were similar, with only slight removal of toxicity after 5 and 10% Ambersorb addition, and partial removal with 10 and 15% PCC addition. *H. azteca* are sensitive to concentrations of PCC greater than 15%, so further PCC tests were not conducted. In Test 4, the addition of 15% Ambersorb significantly reduced toxicity, and the toxicity was recovered in the Ambersorb eluate. A final test was conducted where nonylphenol was measured in sediment and Ambersorb eluate. 10% Ambersorb addition increased survival to 50%, and the resin eluate was toxic. Sediment concentrations of nonylphenol in Tests 3 and 4 were approximately 30-40% of the nominal spiked concentration, but sediment from Test 5 contained 535 mg/kg nonylphenol (Table B-23). The concentration of nonylphenol in the Ambersorb eluate from Test 5 was 93 mg/L, well above the water-only LC50 of 20.7 μg/L (Brooke 1993).

Interstitial Water

The first nonylphenol-spiked IW TIE had 11.4 TU in the baseline test (Table B-24). Filtration reduced the toxicity to 3.8 TU and aeration slightly reduced the toxicity to 9.3 TU. The C18 column did not further reduce the toxicity beyond that of the filtration treatment, and the C18 eluate was not toxic. The second interstitial water TIE evaluated the HLB column (Table B-25). The baseline test contained 8 TU and 1030 μg/L nonylphenol. The HLB column reduced the toxicity to <1 TU and 49.7 μg/L nonylphenol and the column eluate returned 8 TU and 1040 μg/L nonylphenol to clean dilution water. The HLB column was more effective at reducing and returning nonylphenol toxicity than the C18 column.

Freshwater TIE Method Evaluation - Nonylphenol

Solid-phase and IW TIEs were successful at characterizing and identifying nonylphenol toxicity. Ambersorb and PCC reduced toxicity, and the Ambersorb elution successfully recovered toxicity and nonylphenol. The HLB column performed a similar function with the IW sample.

Tetrachlorobenzene

Solid Phase

Three freshwater solid-phase TIEs were conducted with tetrachlorobenzene-spiked sediment (Table B-26). In Test 1, the addition of 5% Ambersorb increased the survival to 64%, while the 20% PCC treatment was toxic. At the time this test was conducted, it was not known that 15% PCC was the upper limit of *H. azteca* tolerance. The Ambersorb elution treatment caused complete mortality. Toxicity in the second solid-phase TIE was reduced by the addition of 5 and 10% Ambersorb and the addition of 15% PCC. Ambersorb was more effective at reducing toxicity than PCC, and significant toxicity was eluted and returned to clean dilution water. In Test 3, the 10% Ambersorb addition significantly reduced toxicity and the toxicity was recovered in the eluate.

Sediments from Tests 2 and 3 were analyzed for tetrachlorobenzene. The concentration in Test 2 was 3563 mg/kg, or 89% of the nominal concentration. The concentration in Test 3 was only 473 mg/kg. Despite the differences in measured concentration, amphipod survival in both baseline tests was the same. Elution of the Ambersorb from Test 3 returned 99.2 mg/L, enough to cause complete mortality.

Interstitial Water

The water-only LC50 for *H. azteca* in tetrachlorobenzene was not derived during this study, and no data on tetrachlorobenzene toxicity to this species was discovered in the literature. The measured concentration in the first interstitial water sample was 1130 µg/L, and produced 1.7 TU (Table B-27), but the measured concentration in the second sample was 2042 µg/L, and did not elicit a toxic response (Table B-28). Unlike the experiment with *E. estuarius*, filtration and aeration did not reduce toxicity of tetrachlorobenzene in IW. The HLB column was ineffective at reducing toxicity in the first IW experiment, but the HLB eluate was slightly toxic. The HLB column removed almost all of the measured tetrachlorobenzene, and returned approximately 10% (Table B-28).

Freshwater TIE Method Evaluation - Tetrachlorobenzene

As in the *H. azteca* experiments with fluoranthene and nonylphenol, the addition of Ambersorb successfully reduced and returned tetrachlorobenzene toxicity. Coupled with PCC in a weight-of-evidence approach, the results using Ambersorb addition to tetrachlorobenzene-spiked sediment indicate the ability to successfully characterize it as the cause of sediment toxicity. Because of the lack of toxic signal in the second interstitial water experiment, and the lack of corroborating results between the experiments, the researchers are unsure if the current TIE treatments would be successful at characterizing or identifying interstitial water toxicity of tetrachlorobenzene to freshwater amphipods.

3.2 Multiple Chemical Toxicity Identification Evaluations

In the following sections, the results for each chemical mixture TIE are presented, followed by a summary discussion of the observed effectiveness of TIE treatments for those mixtures. The TIE results from each mixture are organized into two or three tables. Results from the multiple-chemical solid-phase TIEs are presented in the first table. The interstitial water TIE results are presented in the second and, if necessary, third tables that included chemical confirmation. All applicable tables are provided in Appendix B.

3.2.1 Marine Sediment and Interstitial Water Mixture TIEs

Copper and Fluoranthene

Solid Phase

Two mixture experiments were conducted in which copper and fluoranthene were spiked at 2000 mg/kg and 100 mg/kg, respectively, the same nominal concentrations as in their single chemical tests with marine sediments. Ammonia, nonylphenol, and tetrachlorobenzene were added at one quarter of their nominal single chemical LC50s. In the first experiment, measured concentrations of copper and fluoranthene in the solid-phase mixture were 741 mg/kg and 31.3 mg/kg, respectively, approximately one-third their nominal concentrations (Table B-29). SIR-300 and Ambersorb reduced toxicity, but zeolite did not (Table B-29). A combination of all amendments (SIR-300, Ambersorb, and Zeolite) increased survival to 92%. Sample dilution to 40% (mimicking addition of the three amendments) had little effect on toxicity.

In the second experiment, baseline copper was measured at 1550 mg/kg, compared to the nominal concentration of 2000 mg/kg. Addition of SIR-300 increased survival in solid-phase sediment to 92%. The SIR-300 elution from this mixture returned 174 mg/L copper and a significant amount of toxicity. This is in contrast to the single chemical copper TIE, in which the SIR-300 elution contained 112 mg/L copper and no toxicity (Table B-3).

Fluoranthene was measured at 28.8 mg/kg in the second test solid-phase baseline, approximately one-third the nominal concentration. Addition of Ambersorb increased solid-phase survival to 68% (Table B-30). The Ambersorb eluate contained 180 µg/L fluoranthene, and this eluate was significantly toxic (40% survival). Both fluoranthene and copper eluate concentrations were higher than their respective *E. estuarius* LC50 values.

Interstitial Water

Interstitial water from the first mixture experiment contained 29.5 mg/L copper and 22.3 µg/L fluoranthene. Neither of these concentrations was greater than their respective LC50 values, but there were still 7.9 TU in the baseline test (Table B-31). Filtration reduced the toxicity to 2.8 TU, presumably by sorbing fluoranthene. EDTA reduced the toxicity to 2.4 TU, presumably through chelation of copper. The cation column reduced the toxicity to 1.8 TU. Although elution of the cation column did not appear to result in the same bioavailability artifact as the SIR-300 elution, there was only 1 TU of copper in the interstitial water. This was insufficient to cause toxicity when eluted from the cation column. The HLB column also reduced and returned toxicity, but the sequence of extracting the sample first with the cation column followed by the HLB column did not remove toxicity beyond that removed by the cation column alone. Both of the column eluates were toxic.

The second IW TIE was designed to measure toxicity and quantify extraction and elution of copper and fluoranthene. The baseline test contained 3.6 TU, 97.4 mg/L copper, and 142.9 µg/L fluoranthene (Table B-32). The cation column reduced the toxicity and the copper by half, but the cation eluate only returned 28.1 mg/L, below the *E. estuarius* LC50 of 33.3 mg/L. The cation column also removed fluoranthene, but no fluoranthene was detected in the cation eluate. The HLB column removed fluoranthene, but no fluoranthene was detected in the HLB eluate.

The HLB column reduced the copper concentration by approximately 20%, but no copper was detected in the HLB eluate.

The IW TIEs with the copper-fluoranthene mixtures had variable results. When filtration was used as a pre-treatment for the columns, 65% less toxicity was observed. This masked any further reduction of toxicity by the column treatments. EDTA addition and cation column extraction both reduced toxicity, suggesting metal toxicity, but the cation column eluate did not return toxicity. By itself the HLB column did not remove toxicity, and did not further remove toxicity when used after the cation column in the sequential treatment, yet in both treatments the HLB eluates returned toxicity, indicating an organic contaminant. The sample in the second IW TIE was not filtered. This experiment also demonstrated a stronger reduction of toxicity by the cation column than the HLB column, and weak or no return of toxicity in either of the column eluates. The concentration of copper was reduced by approximately half in the cation column rinsate, but only about half of that concentration was returned in the column eluate. The cation column and the HLB column also reduced the concentration of fluoranthene. Although previous experiments with fluoranthene demonstrated a significant return of toxicity in the HLB eluate, neither column eluate returned toxicity or fluoranthene in this mixture experiment.

Marine TIE Method Evaluation - Copper and Fluoranthene

The solid-phase TIEs were successful at characterizing and identifying copper and fluoranthene toxicity. Individual solid-phase treatments partially reduced toxicity and the combined treatments completely removed toxicity. The eluates from the individual treatment amendments added back significant toxicity and concentrations of chemicals greater than the LC50s. This removal and recapture of the causative chemicals provided useful lines of evidence for toxicant identification. The interstitial water methods produced variable results. The EDTA and cation column treatments were successful at removing toxicity, but toxicity was not recovered in the cation column eluate. Conversely, the HLB column did not reduce toxicity beyond that of the filtration pre-treatment, but toxicity was recovered in some of the HLB eluates. The sequential cation HLB treatment provided evidence of multiple contaminants.

Copper and Ammonia

Solid Phase

Copper and ammonia were spiked at 2000 mg/kg and 600 mg/kg, respectively. Complete mortality was observed in the baseline test. The baseline sediment measured 1985 mg/kg copper, the baseline overlying water measured 55.4 mg/L total ammonia, and 0.538 mg/L unionized ammonia (Table B-33). The overlying water unionized ammonia concentration was less than half the LC50 and was below the U.S. EPA toxicity test protocol application limit of 0.8 mg/L. This suggests minimal contribution to toxicity by ammonia in this sample. This was demonstrated by the reduction of toxicity with the addition of SIR-300 (survival = 88%) with no reduction of total or unionized ammonia. Ambersorb did not reduce toxicity. Zeolite decreased toxicity moderately, and reduced the overlying water total ammonia concentration by approximately 20 mg/L (Table B-33).

In the second solid-phase TIE, the baseline sediment contained 1470 mg/kg copper, and the baseline overlying water contained 52.9 mg/L total ammonia, and 0.844 mg/L unionized ammonia (Table B-34). As in the first solid-phase TIE, the overlying water concentration of unionized ammonia in the baseline was below the LC50, but above the U.S. EPA protocol application limit. The addition of SIR-300 significantly reduced toxicity, but not total ammonia, implying that toxicity was primarily caused by copper. The addition of zeolite reduced ammonia concentrations by approximately 40%, but only slightly reduced toxicity. The reduction of toxicity by zeolite could be attributed to dilution. When eluate from the SIR-300 was added to clean dilution water, the resulting concentration was 42.1 mg/L copper. Although this concentration is greater than the *E. estuarius* copper LC50, the eluate treatment was not significantly toxic. Two previous elution experiments with *E. estuarius* demonstrated the threshold of copper toxicity in SIR-300 eluates is between 112 mg/L and 174 mg/L copper, much higher than the LC50 of 33.3 mg/L that was derived by spiking copper into clean seawater. It is possible that the copper was binding to some component of the resin and had limited bioavailability.

Interstitial Water

The baseline test of the first IW TIE contained 33.8 mg/L copper, 492 mg/L total ammonia, 0.304 mg/L unionized ammonia, and 12.0 TU (Table B-35). The zeolite column reduced the ammonia concentration by 74%. There was low survival in the separate Zeolite blank, for unknown reasons. Sequential treatments were performed in the order listed. Blank survival in the sequential zeolite-EDTA and zeolite-cation column treatments was acceptable. Both combination treatments reduced total ammonia by approximately 70% and toxicity by approximately 77%. The zeolite-cation eluate returned 1.9 TU. The sequential air strip/EDTA and air strip/cation column treatments reduced toxicity to a level similar to that of the sequential zeolite treatments, but did not remove as much ammonia. The air strip-cation eluate only returned a slight amount of toxicity to clean dilution water.

In the second interstitial water TIE, the baseline test contained 35.4 mg/L copper, slightly higher than the LC50. The baseline also contained 456 mg/L total ammonia, 0.549 mg/L unionized ammonia, and 8.0 TU (Table B-36). The cation column treatment reduced the concentration of copper to 9.1 mg/L and the toxicity to 3.6 TU, and did not reduce the total ammonia concentration. The cation column treatment increased the pH, which increased the unionized ammonia concentration to 2.545 mg/L. The cation eluate returned 14.1 mg/L copper to clean dilution water, but this was insufficient to cause toxicity. The zeolite column reduced toxicity to 2.8 TU and total ammonia to 126 mg/L, but also reduced copper to 22.4 mg/L. Burgess et al. (2004) reported that zeolite removed 100% of the spiked copper in their water samples (200 μg/L).

Because ammonia was spiked at a lower concentration than intended, it appears that the SIR-300 and zeolite solid-phase treatments were not able to differentiate between ammonia and copper toxicity. In both solid-phase experiments the ammonia concentration was reduced by the individual addition of zeolite, but based on results reported by Kesraoui-Ouki et al. (1994) and Burgess et al. (2003), the zeolite could have reduced toxicity through the binding of copper. In the individual ammonia spiked tests zeolite performed well, but in samples with a mixture of contaminants and low concentrations of ammonia, other lines of evidence should be used in addition to the zeolite treatment.

In the first IW experiment the cation column reduced the toxicity by almost 50%, and the zeolite column reduced the total ammonia concentration by almost 75%. Together these treatments reduced toxicity further, providing evidence that toxicity was caused by a combination of metals and ammonia. Copper was measured in the second IW experiment to examine relative extraction efficiencies by zeolite and the cation column. Approximately one-third of the total copper was removed by the zeolite column. The researchers concur with Burgess et al. (2004) and recommend that TIEs be designed to account for the possibility of metal binding to zeolite.

Marine TIE Method Evaluation - Copper and Ammonia

When used together, ammonia and metal reduction treatments can differentiate between the two sources of toxicity, particularly in IW samples. In both solid phase tests, the addition of SIR-300 significantly reduced toxicity, and the addition of zeolite slightly reduced toxicity. The overlying water ammonia concentrations indicated that ammonia was contributing to toxicity, but not strongly. The concentration of copper in the SIR-300 eluate demonstrated a significant amount of copper had been removed from the sediment. Individual and sequential treatments in the IW TIEs, coupled with chemical analyses, also successfully characterized and identified the contributions of copper and ammonia to toxicity.

Nonylphenol and Ammonia

Solid Phase

Nonylphenol and ammonia were spiked at 250 mg/kg and 600 mg/kg, respectively. Complete mortality was observed in the baseline test. The baseline sediment contained 162 mg/kg nonylphenol, the baseline overlying water contained 80.1 mg/L total ammonia, and 1.48 mg/L unionized ammonia (Table B-37). The overlying water unionized ammonia concentration exceeded the U.S. EPA guideline of 0.8 mg/L, and was slightly higher than the LC50 of 1.35 mg/L. Addition of individual sediment amendments did not reduce toxicity, but the addition of zeolite reduced the total ammonia by half. Addition of all three amendments together increased survival to 64%, suggesting toxicity was caused by more than one contaminant. There were no dilution effects on toxicity, but the ammonia was greatly reduced in the 40% dilution control. Survival in the treatment blanks was acceptable.

The second solid-phase TIE included measurements of nonylphenol in the sediment and Ambersorb eluate. The concentration of nonylphenol in the baseline was 405 mg/kg, and the concentration of total ammonia and unionized ammonia in the overlying water was 51.0 mg/L and 0.83 mg/L, respectively. The concentration of unionized ammonia was slightly higher than the U.S. EPA guideline of 0.8 mg/L. Addition of Ambersorb slightly reduced toxicity, but the addition of zeolite did not (Table B-38). Zeolite reduced the ammonia concentrations by approximately 38%, and brought the overlying water unionized concentration below 0.8 mg/L. Elution of the Ambersorb returned toxicity and 38.5 mg/L nonylphenol to clean dilution water.

The addition of SIR-300, Ambersorb, or zeolite individually did not reduce toxicity in the first solid-phase experiment, but the multiple-amendment treatment did reduce toxicity and indicated toxicity was caused by more than one class of chemical. Individually, zeolite reduced the overlying water ammonia concentrations below toxic levels, but the presence of toxicity after using the ammonia-reducing treatment indicated the presence of other contaminants. The Ambersorb eluate treatment and related chemical analysis identified the contribution of nonylphenol to toxicity.

Interstitial Water

The baseline test of the first IW sample contained 1470 μg/L nonylphenol, 490 mg/L total ammonia, 1.82 mg/L unionized ammonia, and 8.6 TU (Table B-39). Filtration reduced toxicity to 5.4 TU, but did not significantly reduce the ammonia concentrations. Passing the sample through a zeolite column reduced the total ammonia to 120 mg/L and the unionized ammonia to 0.95 mg/L, less than the *E. estuarius* LC50. The HLB column reduced toxicity to 2.1 TU, but did not reduce ammonia concentrations. The sequential treatment of zeolite and HLB column reduced ammonia and overall toxicity to 1.6 TU. The HLB eluate from this treatment returned 1.8 TU to clean dilution water. Although the sequential air stripping-HLB column treatment did not appear to reduce ammonia concentrations, this treatment was the most successful at reducing toxicity.

The second IW TIE included measurements of nonylphenol in the treatments. Individually, the zeolite and HLB column each reduced toxicity from 8.0 TU to 2.8 TU (Table B-40). The baseline contained 2360 μg/L nonylphenol. The HLB column removed all of the nonylphenol and the zeolite column removed approximately 70% of the chemical. The zeolite column significantly reduced ammonia concentrations below the *E. estuarius* LC50. Elution of the HLB column returned 6.4 TU and 1300 μg/L nonylphenol to clean dilution water.

Marine TIE Method Evaluation - Nonylphenol and Ammonia

Using the solid-phase treatments in combination demonstrated that toxicity was caused by multiple contaminants. Individually, the treatments did not reduce toxicity, but the zeolite reduced ammonia below toxic levels. The presence of toxicity after ammonia reduction indicated it was caused by other contaminants. The elution of the Ambersorb and analysis of the eluate identified nonylphenol. Because the toxic signal was strong enough to overwhelm the individual treatments, sample dilution or an increase in the amount of sediment amendments might have been viable options for increasing the resolution of the treatments.

In these mixture experiments, the IW TIE treatments were more successful at differentiating between the two contaminants. Individually, zeolite and the HLB column reduced toxicity, but when the HLB column was used sequentially with the ammonia-reducing treatments, toxicity was further reduced. As with the solid-phase Ambersorb elution, the return of toxicity and toxic concentration of nonylphenol after elution of the HLB column demonstrates the presence of another contaminant in addition to ammonia.

3.2.2 Freshwater Sediment and Interstitial Water Mixture TIEs

Copper and Fluoranthene

Solid Phase

In the first solid-phase TIE, copper and fluoranthene were spiked at the same concentrations as in the single chemical experiments, 1000 mg/kg and 200 mg/kg, respectively, which resulted in measured concentrations of 1040 mg/kg copper and 121 mg/kg fluoranthene (Table B-41). The other three chemicals were spiked at one-quarter their respective LC50s. Complete mortality of *H. azteca* was observed in the baseline sediment. SIR-300 reduced toxicity, but Ambersorb and zeolite did not. A combination of all three amendments increased survival to 56%. There were no dilution effects and all blanks had acceptable survival.

In the second solid-phase TIE, there was 0% survival in the baseline sediment, and this toxicity was reduced only slightly by addition of SIR-300 (Table B-42). The SIR-300 elution, however, returned 53 mg/L copper, a concentration higher than the LC50 (31 μg/L; Phipps et al., 1995), and this eluate was highly toxic (0% survival). Unlike the SIR-300 elutions produced in the marine TIE, there was apparently no reduction in bioavailability resulting from the elution process. The results were complicated by the observation of 27% survival in the SIR-300 eluate control.

Ambersorb addition only partially reduced sediment toxicity (Table B-42), but the Ambersorb eluate was highly toxic and contained 518 μg/L of fluoranthene, a concentration far higher that the LC50 of 44.9 μg/L (Suedel et al., 1993). Although there was only slight to moderate reduction in solid-phase toxicity with addition of SIR-300 and Ambersorb, because each treatment was overwhelmed by the other contaminant, eluates from both amendments recovered toxic concentrations of the spiked chemicals.

Interstitial Water

The IW from the first mixture experiment contained 18.8 mg/L copper and 179 μg/L fluoranthene (Table B-43). Both concentrations were greater than the water-only LC50 values for *H. azteca*, and the sample was highly toxic with >20 TU. A higher TU value could not be calculated because complete mortality was observed in the lowest concentration tested. Filtration did not remove any toxicity, but EDTA addition reduced toxicity to 7.2 TU, and the cation column reduced toxicity to 12.8 TU. Elution of the cation column returned >20 TU to clean dilution water. The HLB column did not reduce toxicity, but the HLB eluate returned 7.2 TU. When conducted in sequence, the cation and HLB column reduced toxicity to 3.5 TU. Eluates of both columns recovered significant toxicity.

The baseline sample in the second IW TIE contained 16.1 mg/L copper, 142 μg/L fluoranthene, and 5.3 TU (Table B-44). After passing through the cation column, the IW rinsate contained 1.7 mg/L copper, a concentration that greatly exceeded the LC50 of 31 μg/L (Phipps et al., 1995). This post-column rinsate contained 4.6 TU. The cation column eluate had 10.8 mg/L copper and 7.5 TU. The cation column also reduced the concentration of fluoranthene, and 0.7 μg/L fluoranthene was measured in the cation eluate. The HLB column completely removed the fluoranthene, but because 13.0 mg/L copper was still present, the toxicity was only reduced to 4.8 TU. The HLB eluate did not recover any fluoranthene, but there was enough copper in the eluate to cause significant toxicity.

Freshwater TIE Method Evaluation - Copper and Fluoranthene

Though the magnitude of toxicity reduction varied, SIR-300 reduced toxicity in both solid-phase TIEs, and provided a line of evidence implicating metals. There was a similar reduction of toxicity when SIR-300 was combined with Ambersorb, suggesting the metal was the more dominant contaminant. In the second TIE, Ambersorb reduced toxicity. Copper was recovered in the SIR-300 eluate, and fluoranthene was recovered in the Ambersorb eluate, and both eluates recovered toxicity. The elution of these solid-phase amendments provided useful evidence for identifying the toxicants.

Using the cation and HLB columns in sequence produced useful lines of evidence to identify the two primary IW toxicants. In the first IW TIE, EDTA and the cation column partially reduced toxicity, but the HLB column did not. Eluates from both columns returned toxicity, implicating both an organic and a metal. When the cation and HLB columns were used in sequence, there was a greater reduction of toxicity than when they were used separately, and eluates from both columns recovered toxicity. The elution of sequential columns provided additional evidence for toxicity due to both organics and metals. The second interstitial water experiment corroborated these results. Removal and return of toxicity by each column was measured, but chemical analysis of the rinsates and eluates suggest the cation column also removed a large portion of the fluoranthene. Burgess et al. (1997) suggested that the cation exchange column be used as a line of evidence in a Phase II TIE where metal toxicity is already suspected because of the likelihood of high K_{ow} pollutants adsorbing to the column media. Detection of 0.7 μg/L fluoranthene in the cation column eluate confirmed that this PAH had adsorbed to the column. Taken together, the two interstitial water experiments successfully characterized metal and organic toxicity, and analysis of copper and fluoranthene identified the contaminants. Greater sediment dilution or increased amendment masses might have provided more resolution of the relative roles of copper and fluoranthene in these experiments.

Copper and Ammonia

Solid Phase

Copper and ammonia were spiked into sediment at 1000 mg/kg and 1500 mg/kg, respectively. Complete mortality was observed in the baseline test. The baseline sediment contained 992 mg/kg copper, and the baseline overlying water contained 127 mg/L total ammonia, and 5.7 mg/L unionized ammonia (Table B-45). Individually, none of the amendments reduced toxicity, but the addition of zeolite decreased the total ammonia by 85% and lowered the concentration of unionized ammonia below the water-only LC50 of 5.44 mg/L. The addition of SIR-300 also reduced the total ammonia concentration by 43%, but the unionized concentration remained above the LC50. Addition of both SIR-300 and zeolite reduced some toxicity (62% survival).

In the second solid-phase experiment, as expected, separate additions of SIR-300 and zeolite did not reduce toxicity (Table B-46). The zeolite reduced the total ammonia by 71% and the SIR-300 reduced it by 50%. The zeolite treatment reduced the overlying water unionized ammonia below the LC50. The concentration of copper in the sediment was 797 mg/kg and the elution of the SIR-300 returned 13.0 mg/L copper and caused complete mortality when spiked into water. The solid-phase treatment did not identify copper as contributing to toxicity, but the elution of the SIR-300 provided concrete evidence of copper toxicity.

Interstitial Water

The concentration of copper in the first IW sample was 11.6 mg/L. The total ammonia concentration in this IW was 1560 mg/L, the unionized concentration was 6.09 mg/L, and the sample produced >20 TU (Table B-47). Individually, zeolite reduced the total ammonia concentration by 50%, but the unionized concentration was 5.38 mg/L, and the sample remained toxic. The cation column did not reduce the ammonia concentration, but reduced toxicity to 7.5 TU. When ammonia reduction treatments were used sequentially with metal reduction treatments, the toxicity was greatly reduced. EDTA combined with zeolite or air stripping reduced the toxicity to 3.0 and 2.8 TU, respectively. The cation column combined with zeolite and air stripping reduced toxicity to 2.9 and 4.0 TU, respectively. The cation eluates from both sequential treatments returned significant toxicity, but the cation column conducted in sequence with air-stripping was more efficient at returning toxicity in the eluate. This increased efficiency might have been caused by the binding of copper in the zeolite column.

In the second interstitial water experiment, toxicity was not reduced by either the cation column or the zeolite column, but chemical analysis indicated removal of 98% of the copper and 64% of the ammonia by the respective columns (Table B-48). The cation column removed most of the copper, but the ammonia concentration was sufficient to cause toxicity. Similarly, the zeolite column removed 64% of the ammonia, and 60% of the copper, but toxic concentrations of both constituents were still present.

Freshwater TIE Method Evaluation - Copper and Ammonia

In the solid-phase TIEs, toxicity was reduced when SIR-300 and zeolite were used in combination, even though toxicity was not reduced when either was added separately. The use of combined amendments is useful for identifying toxicants in mixtures. Eluate analysis was very useful; even when SIR-300 did not reduce toxicity (because of the presence of ammonia), the SIR-300 eluate returned enough copper to implicate the metal contributed to toxicity. Sequential column treatments in the IW TIEs differentiated between copper and ammonia toxicity.

The SIR-300, which is used to treat metals, also reduced total ammonia, but actually increased the unionized ammonia fraction. Zeolite, which is used to treat ammonia, reduced copper concentrations in the IW test. Burgess et al. (2003) found that zeolite addition reduced bioavailability of metals in marine TIEs, but Besser et al. (1998) showed that the addition of zeolite to sediments containing high concentrations of cadmium and copper did not reduce acute toxicity or concentrations of the metals in fresh interstitial water. Because zeolite, SIR-300, Ambersorb, and powdered coconut charcoal each may bind chemical classes other than those for which they are intended, it is important to be aware of this capacity when considering the results. Removal of toxicity through binding of metals and organic chemicals is confirmed through Phase II TIE solvent elution procedures. In the case of SIR-300 and Ambersorb, this involves resin elutions using acid and acetone, respectively. The potential for misinterpretation of results in TIEs involving zeolite can be minimized by considering results of the other TIE treatments that target ammonia (e.g., pH adjustments, air stripping).

Nonylphenol and Ammonia

Solid Phase

In the first solid-phase TIE, nonylphenol and ammonia were spiked at 800 mg/kg and 1500 mg/kg, respectively. Complete mortality was observed in the baseline test. The baseline sediment contained 74 mg/kg nonylphenol, the baseline overlying water contained 143 mg/L total ammonia, and 4.96 mg/L unionized ammonia (Table B-49). The overlying water unionized ammonia concentration was below the LC50 of 5.44 mg/L. Individual amendments did not reduce toxicity, but the combination of amendments increased survival to 48%. The individual zeolite treatment reduced overlying water ammonia by 77% and lowered the unionized ammonia concentration below the LC50. Ambersorb also reduced overlying water ammonia, but this was probably due to dilution.

Similar results were observed in the second solid-phase TIE, but the Ambersorb increased survival by 14% (Table B-50). Total and unionized ammonia concentrations were lower in this experiment and this apparently allowed the Ambersorb to reduce toxicity. The sediment contained 525 mg/kg nonylphenol and the Ambersorb elution returned 62 mg/L nonylphenol, which caused complete mortality.

Interstitial Water

The first IW TIE contained 714 μg/L nonylphenol, 860 mg/L total ammonia, 6.67 mg/L unionized ammonia, and 12.0 TU (Table B-51). Individually, the zeolite column and the HLB column partially reduced toxicity to 1.9 and 8.6 TU, respectively. The zeolite column reduced the total ammonia concentration by 75% and reduced the unionized ammonia concentration below the LC50. When used sequentially with the ammonia-reducing treatments, the HLB column further reduced toxicity and the elution of the HLB column successfully returned toxicity to clean dilution water. In the sequential treatments, the zeolite column reduced total ammonia by 55%, and the air stripping reduced the total ammonia by only 23%. Neither treatment reduced sufficient ammonia to reduce the unionized concentration below the LC50. The sequential zeolite-HLB column was the most effective at returning toxicity, when the HLB column was eluted and the eluate was spiked into clean dilution water.

The baseline sample of the second IW experiment contained 8.0 TU, 1050 mg/L total ammonia, and 8630 µg/L nonylphenol (Table B-52). The HLB column reduced toxicity to 6.1 TU, and the concentration of nonylphenol to 76.9 µg/L, but also reduced the total ammonia to 410 mg/L, which had not previously occurred with the HLB column. The HLB eluate returned 1370 µg/L nonylphenol to clean dilution water, and caused complete mortality. The zeolite column significantly reduced ammonia, and removed over 90% of the nonylphenol, but did not reduce toxicity.

Freshwater TIE Method Evaluation - Nonylphenol and Ammonia

Addition of individual sediment amendments did not reduce toxicity, but the addition of zeolite reduced the total ammonia by half. Addition of all three amendments together increased survival to 64%. These results indicate that toxicity was caused by more than one contaminant. There were no dilution effects on toxicity, but the ammonia was greatly reduced in the 40% dilution control. Survival in the treatment blanks was acceptable.

The IW TIE treatments were able to differentiate between the two contaminants. Individually, zeolite and the HLB column reduced toxicity, but when the HLB column was used sequentially with the ammonia-reducing treatments, toxicity was further reduced. Toxicity was returned with both HLB column eluates. In the second IW TIE, chemical analysis demonstrated incomplete return of nonylphenol in the HLB eluate and significant removal of nonylphenol by the zeolite column.

3.3 Ambient Sample TIEs

Six ambient sample TIEs were conducted as a final evaluation of the sediment TIE methods. Three marine and three freshwater sediments were collected from four sites recommended by the PSC, and from two additional sites in California. The three marine sites were located in southern California: Switzer Creek (SZ), Upper Newport Bay (UN) and Consolidated Slip (CS). The three freshwater sites were San Diego Creek (SC) in southern California, Alisal Creek (AC) in central California, and Indiana Harbor (IH). Detailed toxicity results of the six ambient sample TIEs are presented in Appendix B (Tables B53 to B81) and summarized below in Tables 3-1 to 3-6. Sediment chemistry was measured on undiluted sediment and interstitial water, or on dilutions of the interstitial water, where appropriate. Selected chemistry of diluted sediment and interstitial water samples are also presented in Appendix B, and complete chemical analyses are presented in Appendix C. Final interpretations of the ambient sample TIEs are summarized in Table 3-7. Table 3-7 also describes which of the three TIE phases were successfully completed: Phase I (characterization), Phase II (identification), or Phase III (confirmation). Phase I TIEs are designed to characterize whether the cause of toxicity is due to ammonia, metals, or organics, or a combination of these three classes of toxicants. During Phase II, the TIE procedures are combined with chemical analyses to identify specific chemicals responsible for toxicity. This includes addition of resins to reduce toxicity of sediments and the solid-phase extraction columns to treat interstitial water. Additional Phase II steps include elution of resins or solid-phase extraction columns with appropriate solvents, and by toxicity tests of the eluates. Phase III confirmation may involve a number of approaches. This research emphasized the species sensitivity approach where concentrations of chemicals measured in different test matrices are compared to known effects threshold derived using experimental or empirical procedures. In the TIEs described below, lines of evidence are derived from all three phases of the TIE process.

To provide information on the relative sensitivity of the amphipods *H. azteca* and *E. estuarius* to chemicals identified in the ambient samples, chemistry data were evaluated based on published sediment LC50 and other effect values for these species, as well as freshwater and marine sediment quality guideline values. Effect values that were derived from spiked sediment dose-response experiments were emphasized in this evaluation, including unpublished values that were derived at MPSL and other laboratories. Published guidelines values were used for chemicals lacking dose-response data. Evaluation concentrations are listed in the individual ambient chemistry summary tables in Appendix B. The Effects Range Median guidelines were used for evaluating the marine ambient samples (Long et al., 1995), and the consensus-based Probable Effects Concentration guidelines were used to evaluate the freshwater ambient samples (MacDonald et al., 2000b). Polychlorinated biphenyls were evaluated using the consensus-based values of MacDonald et al. (2000a). Polycyclic aromatic hydrocarbons (PAHs) were evaluated using sediment quality guidelines, the organic-carbon normalized Total PAH Median Effect Concentration (MEC) of Swartz (1999), and the target lipid model (DiToro and McGrath, 2000; Joy McGrath, HydroQual Inc., Mahwah, NJ, U.S., personal communication). The concentrations of PAHs in the marine samples were also summed based on Page et al. (2002) and compared to a medial lethal concentration for the marine amphipod *Rhepoxynius abronius*. Interstitial concentrations of ammonia and sulfide were measured at the initiation and termination of the initial sediment tests. With the exception of ammonia in the Indiana Harbor sediment, all other overlying water and interstitial concentrations of ammonia and sulfide were below tolerance thresholds for the test organisms. Grain size results were also evaluated (Table C-9). The concentrations of fine grained material were within the tolerance limits of the test organisms.

3.3.1 Switzer Creek

Switzer Creek (SZ) is a marine station located in San Diego Bay, in southern California, and this site is the focus of an ongoing sediment TMDL study due to sediment pollution, toxicity and benthic community impacts. The station is located at the mouth of the Switzer Creek drainage where it enters the bay. Switzer Creek lies in the Pueblo Watershed which encompasses 60 square miles of the most densely populated area of San Diego County. Major development in the watershed is comprised of residential (45%), retail and office (11%) and industrial (10%) land uses. Previous studies of Switzer Creek sediments have shown that sediment toxicity and contamination at this site are highly seasonal, with the greatest toxicity and contamination observed during the rainy season (Anderson et al., 2004).

Solid-Phase TIEs

The initial solid-phase test with Switzer Creek sediment was conducted using 25, 50, 75, and 100% concentrations using sediment that had been dry-sieved through a 280 μm screen to aid in the separation of the TIE amendments (Table B-53). All concentrations were significantly toxic. The initial test with interstitial water (IW) extracted from this sediment was significantly toxic at 100% concentration (20% survival), and 90% survival was observed in the 50% IW concentration. This was considered insufficient toxicity to conduct an IW TIE, and solid-phase procedures were emphasized in subsequent experiments. Results of the first solid phase test demonstrated that SZ sediment diluted to 25% resulted in complete mortality. This was considered sufficient toxicity to allow resolution between the TIE treatments. An additional solid-phase dilution series was performed to determine the optimal concentration for measuring changes in amphipod tissue concentrations after application of the different solid-phase TIE treatments (described below).

Two solid-phase TIEs were performed in which SIR-300 and Ambersorb were added to 25% SZ sediment (Table B-54). Survival in the baseline tests was 12% and 0% in TIEs 1 and 2, respectively. Toxicity in both experiments was reduced with addition of SIR-300 and Ambersorb. Addition of SIR-300 was less effective than Ambersorb in the second experiment. The resins were recovered from the sediment via sieving, and were eluted in both experiments. Ambersorb was eluted with acetone and SIR-300 was eluted with hydrochloric acid. Significant toxicity was observed in water spiked with the Ambersorb eluate, but not in water spiked with the SIR-300 eluate. Additional Ambersorb from the second experiment was eluted for further Phase II High Pressure Liquid Chromatography (HPLC) fractionation. Aquascience (Davis, CA) performed a standard Phase II fractionation gradient using 30-100% methanol (U.S. EPA 1993). This resulted in 30, one-mL methanol fractions. A portion of these fractions (300 μL) was reconstituted in clean dilution water, and exposures with *E. estuarius* were conducted. The only response observed was complete mortality in the 27th fraction (data not shown). Organochlorine pesticides, pyrethroid pesticides and PAHs were measured in this fraction using GC-MS.

Solid-Phase Chemistry

Chemicals were measured in undiluted SZ sediment. To approximate the concentrations of chemicals in the sediment dilution used in the TIEs (i.e., 25% sediment), results of the analyses were divided by four. Both measured and calculated chemical concentrations were evaluated for their contribution to sediment toxicity.

The pyrethroid pesticides bifenthrin, cyfluthrin, and permethrin were detected in undiluted SZ sediment. Concentrations of bifenthrin and permethrin in undiluted sediment were above the sediment LC50 values for *E. estuarius* (MPSL unpublished data). The concentrations of bifenthrin and permethrin did not exceed the sediment LC50s in diluted sediment. There is no *E. estuarius* sediment LC50 for cyfluthrin, but the concentrations of cyfluthrin in diluted and undiluted sediment exceeded the *H. azteca* LC50s (13.4 TU; Amweg et al., 2005). Because *H. azteca* and *E. estuarius* have similar sensitivities to bifenthrin and permethrin, it was surmised that the concentration of cyfluthrin in the 25% SZ sediment was sufficient to account for *E. estuarius* mortality.

The total chlordane concentration measured in undiluted SZ sediment was 62.6 ng/g, and the concentration of chlordane estimated to be in the 25% SZ sample was 15.7 ng/g. Nautilus Environmental conducted sediment and water spiking experiments to determine toxicity of chlordane to *E. estuarius*, and found the 10-day No Observed Effect Concentration (NOEC) in solid-phase tests to be greater than 49 ng/g, and the water-only NOEC to be greater than 310 μg/L (personal communication, Chris Stransky, Nautilus Environmental, San Diego, CA). Trace concentrations of DDT were also detected in this sediment but these were well below the published effect threshold for *E. estuarius*. Oxadiazon was also detected in the sediment and subsequent baseline tissue exposure (discussed below), but the *E. estuarius* threshold for oxadiazon is unknown.

Polycyclic aromatic hydrocarbons (PAHs) were evaluated for potential toxicity to *E. estuarius* using methods described by Page et al. (2002). This approach is similar to the sum PAH model developed by Swartz et al. (1995), but considers toxicity of both the parent PAH compounds and their transformation products. Page et al. (2002) calculate concentrations of total PAHs and compare this to an LC50 developed for the marine amphipod *Rhepoxynius abronius* (LC50 = 10,750 ng/g). Total PAHs in SZ sediment were well below the sediment LC50 for *R. abronius* in both the 100% (6756 ng/g) and 25% (1689 ng/g) sediment dilutions.

The concentrations of trace metals in SZ sediment were generally low. No metals exceeded the Effects Range Median (ERM) guideline values (Long et al., 1995), and the copper concentration was well below the sediment LC50 for *E. estuarius* (Table B-55). Total PCBs were well below the Median Effect Concentration of 400 ng/g (MacDonald et al., 2000a), and organophosphates were not detected in SZ sediment.

After HPLC fractionation, only the 27th methanol fraction was significantly toxic to *E. estuarius*. This fraction was analyzed for organochlorines, PAHs, and pyrethroid pesticides. No PAHs were detected in fraction 27, and the concentration of chlordane in this fraction was 0.085 µg/L (Table B-55), well below the water-only NOEC for *E. estuarius* (NOEC = >310 µg/L; personal communication, Chris Stransky, Nautilus Environmental, San Diego, CA, U.S.). Although relatively high concentrations of pyrethroids were measured in SZ sediment, none were detected during the original analysis. As summarized below in the weight-of-evidence discussion of the SZ TIE results, methods used to measure organic chemicals in the HPLC fraction may have resulted in dilution of the samples. In TIEs of the other ambient sample, the chemists involved in this project used direct injection of the acetone eluates to measure organic chemicals, and in many cases these showed elevated concentrations of pyrethroid pesticides in the acetone eluates. To investigate this possibility in the SZ samples, the researchers conducted additional experiments with Ambersorb, and analyses of the Ambersorb acetone eluates via direct injection in the gas chromatograph are pending.

Tissue Exposure and Analysis

A third solid-phase TIE was conducted with 10% SZ sediment. This TIE included additional replicates to measure effects of the Ambersorb and SIR-300 additions on uptake of selected organics and metals in amphipod tissues. Because our previous results indicated that both SIR-300 and Ambersorb reduced toxicity of SZ sediment, this experiment was designed to provide additional evidence of the relative roles of metals and organics in contributing to the toxicity of this sample. The 10% sediment concentration was selected to cause sufficient toxicity to resolve differences between the TIE treatments, but allow survival of enough amphipods to provide tissue for chemistry. Approximately 50% mortality was anticipated in the baseline sample with 10% SZ sediment. Amphipods were exposed to SZ sediment with and without SIR-300 and Ambersorb. Baseline survival was 46% and both amendments significantly reduced toxicity (Table B-54). As before, the Ambersorb treatment was more effective at reducing toxicity than SIR-300 (96% survival vs. 68% survival).

Tissues in the baseline, SIR-300 and Ambersorb treatments, and the control were analyzed for pyrethroids, organochlorines, PCBs, and metals. Only two organochlorines were detected in tissues. Oxadiazon was detected in the baseline sample (28.7 ng/g fresh weight), and

5.14 ng/g DDE (o,p') was detected in the SIR-300 treatment. No pyrethroids or PCBs were detected in any of the tissues. Metals were detected in the tissues from all four samples, but there were no differences in metal concentrations among the treatments, indicating metals probably did not contribute to toxicity.

Weight of Evidence

Addition of SIR-300 and Ambersorb both reduced toxicity of the 25% SZ sediment sample, but only the elution of Ambersorb returned toxicity to clean dilution water. Results from the Phase I treatments suggested toxicity was caused by metal and organic chemicals. Results of the Phase II resin elutions suggested toxicity was caused by organic chemicals, because no toxicity was observed in the SIR-300 eluate. Additional Phase II treatments included the HPLC fractionation of the Ambersorb eluate. A single toxic fraction (#27) was identified and analyzed for pyrethroids, organochlorines, and PAHs. Analysis of the fraction identified a low concentration of chlordane, but no other compounds. Another experiment with tissue chemistry characterized the cause of toxicity as an organic, but only oxadiazon was detected in the baseline tissues.

The Phase III procedure in this TIE consisted of comparing the chemical concentrations in the various test matrices to known effects thresholds for *E. estuarius*. Chemical concentrations were evaluated in both undiluted and diluted sediment. Using only the LC50 values for *E. estuarius* in Table B-55, there are a total of 4 TU of pyrethroids in the undiluted sediment and 1 TU in the diluted sediment. The number of toxic units increases when the *H. azteca* LC50 values are used to calculate the contribution of cyfluthrin. There are 1.9 toxic units of organic carbon-normalized pyrethroids in the sediment. The concentration of chlordane in the 100% and 25% SZ sample also exceeded its Probable Effects Level (PEL) sediment quality guideline value, and this was the only compound detected in the elution of the Ambersorb. However, dose-response data with this pesticide suggest the concentration of chlordane in the sediment and Ambersorb eluate was below the NOEC for *E. estuarius*.

While some of the TIE methods were useful in characterizing the cause of toxicity in this TIE, it was not possible to conclusively identify the specific cause(s) of toxicity in the Switzer Creek sample. This was primarily due to non-detection of any significant concentration of chemicals in the Ambersorb eluate, despite the fact that it was significantly toxic to amphipods when spiked into clean seawater. Our conversations with the chemists involved in these analyses suggest this might have been due to incomplete recovery of the chemicals in key steps of the TIE and analytical process. In these experiments, the Ambersorb was eluted with acetone then subjected to HPLC fractionation with methanol. It is possible the acetone elution was not completely efficient at solubilizing all organic chemicals sorbed to the resin, and further loss could have occurred during the HPLC process. The HPLC methanol fraction was spiked into water then re-solubilized in iso-octane as a solvent exchange step prior to gas chromatography. The chemists estimate this resulted in an approximately five-fold dilution of the HPLC methanol fraction. The chemical and TIE evidence from the Switzer Creek TIE suggest toxicity was caused by pyrethroid pesticides. These are particularly problematic in the TIE and analytical process because they are highly insoluble, and therefore difficult to recover from the extraction media. In addition, they are very toxic to amphipods, and the detection limits for these compounds are sometimes within the range of their LC50s to the test organisms. As discussed below, future TIEs will use an increasingly non-polar methylene chloride/methanol elution series to more completely extract more non-polar compounds from the Ambersorb and HLB columns.

The research team also is collaborating with the chemists involved in this project to alter procedures to minimize loss of chemicals during the sample preparation steps. The ability to amend the TIE and analytical procedures to address these deficiencies during the current project was constrained by the 12 week turn-around time between the initial sample submittal and receipt of the final GC-MS results. This illustrates the importance of timely analyses and good communication between the TIE and chemistry analysts, particular when new procedures are being developed. Because TIEs are an iterative process, where each step depends on both the toxicity and chemistry results from previous steps, the process works best when the toxicology and chemistry teams are in constant communication regarding the methods being employed, and the results obtained.

As noted above, rather than using a solvent exchange step, the chemists who conducted the Ambersorb eluate analyses used direct injections of the acetone eluates in subsequent TIEs described below. These resulted in detections of toxic concentrations of pyrethroids in some of the other ambient samples. In follow-up experiments with SZ sediment, the researchers conducted additional acetone elutions of Ambersorb, and analyses of these eluates are pending

3.3.2 Consolidated Slip

Consolidated Slip (CS) is a marine station located at the base of the Dominguez Channel where it enters Long Beach Harbor in southern California. The Dominguez Channel drains an approximately 100 square mile watershed located in the intensely urbanized area west of the Los Angeles River in the Los Angeles Basin. Inputs into this part of the system include urban runoff and non-processed industrial waste discharges (Anderson et al., 2001). There are approximately 70 NPDES discharges in the Dominguez Channel Watershed Management Area, and an additional 424 permitted stormwater discharges. Major discharges into the watershed include a secondary treatment POTW, two generating stations, and six refineries (LARWQCB 2004). Sediments from the site have been shown to be consistently toxic in a number of previous studies (Anderson et al., 2001; Moore et al., 2004; Young et al., 2005), and these studies have shown CS sediment to be highly contaminated by contaminant mixtures including PAHs, PCBs, organochlorine pesticides, and metals.

Solid-Phase TIEs

Initial sediment and interstitial water tests with Consolidated Slip (CS) sediment demonstrated it was highly toxic to amphipods (Table B-56). Sediments and interstitial waters from this station were black and oily and had the characteristic odor of hydrocarbons. Complete mortality was observed in sediment diluted to 1%. A solid-phase TIE was conducted on 1% sediment using additions of SIR-300 and Ambersorb individually, and in combination. Survival in the baseline 1% CS sample was 12% (Table B-57). Addition of SIR-300 increased survival to 68%, and addition of Ambersorb increased survival to 100%. Survival in CS sediment treated with both Ambersorb and SIR-300 was also 100%. Survival in the 10 and 20% dilution blanks was 44 and 48%, respectively. These results indicate that part of the reduction of toxicity of the 1% CS sample was caused by dilution due to addition of the resin masses. Survival in the 10% dilution blank was not significantly different from the SIR-300 treatment, indicating that dilution probably played a role in the reduction of toxicity in the SIR-300 treatment.

Both sediment amendments were eluted, but only the Ambersorb acetone eluate was toxic. The acetone extract was fractionated with HPLC, but no toxicity was observed in any of the 30 individual fractions (data not shown). A subsample of the 100% acetone extract was submitted for chemical analysis.

Solid-Phase Chemistry

The following measurements of chemicals in CS sediment are based on analysis of 100% sediment. These results were divided by 100 to estimate the concentrations of chemicals in the sediment dilution used in the solid-phase TIEs (1% sediment). Both measured and estimated chemical concentrations were evaluated for their contribution to sediment toxicity.

Individual polycyclic aromatic hydrocarbons (PAH) compounds did not exceed sediment quality guidelines, and the sum of 13 PAH parent compounds, normalized based on organic carbon in the sediment, did not exceed the consensus-based median effect concentration of 1800 μg/g OC (Swartz 1999). PAHs were also summed based on the method of Page et al. (2002), which considers toxic effects due to parent compounds and their transformation products. Total PAHs exceeded the sediment LC50 for *Rhepoxynius abronius* in both undiluted and diluted sediment (Table B-61, Page et al., 2002). PAH compounds also were summed using the target lipid model (Di Toro et al., 2000; Di Toro and McGrath, 2000; Joy McGrath, HydroQual, Inc., Mahwah, NJ, U.S., personal communication), which sums parent and transformation compounds and calculates species-specific toxic units for each compound. Based on this model, the concentrations of PAHs in undiluted sediment contributed approximately 5 TU, and would be expected to contribute significantly to the cause of toxicity. However, PAHs in 1% CS sediment contributed approximately 0.05 TU and were not expected to cause toxicity based on this model.

The concentrations of trace metals in the sediment were generally low in 1% CS sediment, but several metals in 100% sediment exceeded their ERM values. No metals exceeded ERM guideline values in 1% CS sample, and the copper concentration was well below the sediment LC50 (Table B-61). The concentration of copper was twice the sediment LC50 for *E. estuarius* in undiluted CS sediment (MPSL, this study). Three organochlorines also were above ERM values but the concentration of OC-normalized total DDT was well below the published LC50 value. The concentration of total PCBs in 100% sediment approached the consensus-based median effect concentration (MEC; MacDonald et al., 2000a), but was well below the MEC in 1% sediment. Pyrethroid and organophosphate pesticides were not detected in CS sediment.

Although the original Ambersorb eluate was toxic, none of the HPLC methanol fractions were toxic. The original acetone eluate was analyzed for several classes of organic chemicals. Many PAH compounds were detected, but their concentrations were well below the water-only LC50 values for *E. estuarius*. Although no pyrethroids were detected in the undiluted sediment, some pyrethroids were detected in the Ambersorb eluate when the acetone was directly injected into the gas chromatograph. Water-only LC50s are not available for *E. estuarius*, but concentrations of bifenthrin and permethrin were greater than the water-only LC50s for *H. azteca* which is comparable to E. estuarius in its sensitivity to pyrethroids. Detection of pyrethroids in the eluates without detection in the sediments could be attributed to insufficient sediment method detection limits and/or to the possibility that the Ambersorb concentrated low levels of sediment pyrethroids during the exposure.

Interstitial Water

CS interstitial water caused complete mortality when diluted to 10%. Three IW TIEs were conducted using two separate dilution series. The first TIE was conducted on concentrations ranging from 1 to 10% CS IW (Table B-58). Complete amphipod mortality was observed in the baseline test down to 1% sample. Survival increased to 100% when the IW sample was passed through a glass fiber filter (conducted as a pre-treatment step to remove particles prior to the solid-phase extraction column treatments). The sample had negligible particle load based on visual inspection, and it was assumed that the reduction of toxicity was due to contaminants binding to the filter material. The filter was not eluted with solvent. Because toxicity was removed by the filter treatment, further reduction of toxicity was not observed in the solid-phase extraction column treatments (e.g., cation and HLB columns). Eluate from the HLB column was toxic, indicating an organic contaminant was sorbed to the column.

Because of extreme IW toxicity, the second TIE was conducted on a dilution series ranging from 0.25% to 2% IW (Table B-59). Although IW was extracted from the sediment via centrifugation, minute particles are observed in some samples after the initial centrifugation step. An additional centrifuge treatment was added to remove any residual particles in CS interstitial water prior to solid-phase extraction with the cation and HLB columns. In the second IW TIE, toxicity was reduced from 641 TU to <50 TU by both individual and sequential column treatments (Table B-59). Elution of the cation column did not add back toxicity, but elution of the HLB column with methanol returned more than half the TUs.

The third IW TIE emphasized serial fractionation of the HLB column using increasing concentrations of methanol (U.S. EPA 1993). In this TIE, two HLB columns were used to extract 2% CS IW and these were then eluted with methanol. One column was extracted in the normal manner with 100% methanol. The other was serially extracted with nine fractions which used progressively higher concentrations of methanol. The centrifugation treatment reduced toxicity from 667 TU to 214 TU, and the HLB column further reduced toxicity to <50 TU (Table B-60). Because there were negligible particles observed in the sample, it was assumed that the reduction of toxicity by the centrifuge treatment was due to sorption of chemicals to the surfaces of the polyethylene centrifuge tubes. Elution of the first HLB column returned 71 TU, and elution of the second HLB column with 100% methanol returned 65 TU. These results suggest incomplete elution of contaminants from the HLB column using 100% methanol, because >140 TUs were unaccounted for in both instances. HLB elution fractions containing less than 100% methanol did not return any toxicity to clean dilution water, indicating the contaminants were very hydrophobic.

Interstitial Water Chemistry

Chemical analyses were not conducted on the undiluted CS interstitial water sample, which was black in appearance and smelled strongly of hydrocarbons. Chemicals were measured in 10% CS IW sample, because of the volume of IW necessary for analysis, and because there were concerns about analytical interference from the highly contaminated interstitial water. The results were divided by 5 to estimate the concentrations of chemicals in the IW dilution used in the TIEs (2%). Both measured and estimated chemical concentrations were evaluated for their contribution to toxicity.

Although the interstitial water caused significant toxicity to *E. estuarius*, there were no chemicals at concentrations in 10% IW greater than their respective toxicity thresholds (Table B-61). The concentrations of IW PAHs were well below published LC50 values for *E. estuarius*, but these most likely contributed small, but additive, toxic units. The contribution of PAHs in the IW was evaluated using the target lipid model, but without a measurement of dissolved organic carbon to account for PAH bioavailability, it was assumed that the results overestimated toxicity. There were also low concentrations of p,p'-DDD and p,p'-DDE measured in CS IW. These compounds amounted to approximately one-half TU total DDT in 10% IW. The concentrations of metals were well below published thresholds for *E. estuarius*, and the pyrethroids, PCBs, and organophosphates were not detected in CS IW.

The first HLB column eluate from the TIE conducted on 2% CS interstitial water was analyzed for organics (Table B-61). No organochlorines, PCBs, organophosphates, or pyrethroids were detected in eluate from the HLB column. The only significant concentrations of chemicals measured in the HLB eluate were PAHs, but concentrations of these compounds were below organism thresholds.

Weight of Evidence

Both sediment and interstitial water TIEs were conducted on diluted Consolidated Slip samples and treatments in both matrices reduced toxicity. Ambersorb and SIR-300 reduced solid-phase toxicity and characterized the cause of sediment toxicity as organics and metals. The Phase II elution of the Ambersorb amendment returned toxicity, whereas the elution of the SIR-300 did not. Phase II treatments included the HPLC methanol fractionation of the Ambersorb eluate. The HPLC fractionation method resulted in no toxicity in any of the 30 fractions, so organic chemicals were analyzed in Ambersorb acetone eluate. Concentrations of organics in the sediment and Ambersorb eluate were compared to toxicity thresholds as part of the Phase III evaluation. The concentration of total PAHs in the sediment was evaluated using three methods of summation: Page et al. (2002), Swartz (1999), and the target lipid model (DiToro and McGrath, 2000). The concentration of total PAH in undiluted CS sediment exceeded the thresholds of Page et al., (2002) and the target lipid model. The Page et al. (2002) threshold was also exceeded by total PAHs in diluted sediment. The organic carbon-corrected LC50 of Swartz (1999) was not exceeded, but this summation only considers parent compounds, and in the case of CS sediment, the concentrations of many of the transformation products far exceeded the concentrations of the parent compounds.

The concentrations of PAHs in the Ambersorb eluate were below the threshold values for *E. estuarius*. Although there was significant toxicity in the Ambersorb eluate, no other compounds were detected. The weight of evidence from the solid-phase TIEs and chemistry suggests that PAHs are the cause of toxicity in 1% CS sediment, but the low concentrations of detectable PAHs in the Ambersorb eluate suggest that, while the toxicity signal was adequate, the methods used were inadequate for detecting toxic concentrations of any compounds in the eluate. As discussed in the Switzer Creek TIE, it is suspected that loss of organic chemicals occurred during the Ambersorb elution and solvent exchange steps prior to gas chromatography.

The Phase I IW TIE methods also characterized the cause of toxicity as an organic. Toxicity was reduced by both the cation and HLB SPE columns, but only the HLB column eluate returned toxicity. IW and HLB eluate chemistry were evaluated by comparing measured concentrations to threshold values. As with the Ambersorb eluate, PAHs were detected in the HLB eluate, but concentrations of individual PAHs were below toxicity thresholds. The evaluation of toxicity due to total PAHs using the target lipid model were constrained by a lack of dissolved organic carbon measurement to account for bioavailability. Using a DOC estimation of 50 mg/L (Williamson and Burgess, 2003), the target lipid model was used to estimate 0.5 TU of PAHs in 2% interstitial water. Because the PAH mass in CS interstitial water was comprised largely of PAH transformation products, a model that considers toxicity due to these PAHs is more appropriate than one that considers parent compounds alone. In addition to PAHs, there was approximately one-half TU of total DDT in the IW.

The weight-of-evidence from the CS TIE and chemical analyses suggest PAHs are the likely cause of toxicity, but this is limited by a lack of conclusive identification of chemicals in the Phase II eluates. The research group recommends that amendments to the TIE methods include use of an increasingly non-polar elution series of both the Ambersorb and HLB extraction media to ensure more efficient elution of hydrophobic compounds. In addition, measures of interstitial water dissolved organic carbon are necessary to apply appropriate toxicity models that consider toxicity due to parent PAH compounds and their transformation products. Detections of toxic concentrations of pyrethroid pesticides in the Ambersorb acetone eluates that were subjected to direct injection GC-MS suggest these also could have played a role in toxicity of CS sediments.

3.3.3 Upper Newport Bay

Upper Newport Bay (UN) is an estuary located in central Orange County, in southern California, and is one of the largest remaining wetland areas in this region. The UN station is approximately 0.5 km downstream of the San Diego Creek station (SD) and this channelized creek is the largest contributor of freshwater flow into Upper Newport Bay. This watershed encompasses 154 square miles in this highly urbanized region. Land use in this watershed is comprised of 90% urban, including mixed-use residential with retail businesses and offices, light industry, nurseries, an airport, a university, and military bases. Agriculture accounts for 7% of the remaining land use. Upper Newport Bay is the subject of ongoing TMDL studies for a number of toxic pollutants including organochlorine and organophosphate pesticides.

Solid-Phase TIEs

The first test with Upper Newport Bay (UN) sediment demonstrated complete mortality in undiluted sample (Table B-62). Based on previous knowledge of inputs to this watershed, the solid-phase sample was also treated with a carboxylesterase enzyme to reduce toxicity of pyrethroid pesticides, and with the metabolic inhibitor, piperonyl butoxide (PBO). PBO can either act as a synergist of pyrethroid pesticide toxicity, or can reduce toxicity of metabolically-activated organophosphate pesticides. These two treatments were added to the overlying water in the exposure (see Appendix A). Survival in the baseline sample was 0%, and none of the treatments reduced toxicity, suggesting contaminant concentrations were sufficient to overwhelm the TIE treatments (Table B-63).

A second solid-phase initial test was conducted after the solid-phase treatments in the first TIE were ineffective at reducing toxicity of 100% sample. The second test resulted in 4% survival in 10% sample. In the second solid-phase TIE, the combined SIR-300 and Ambersorb treatment significantly reduced toxicity (Table B-63). Although Ambersorb alone did not reduce toxicity, the amendment was eluted with acetone to investigate whether toxic concentrations of contaminants were bound to the resin. Significant toxicity was observed in the Ambersorb eluate, indicating organic contaminants contributed to toxicity of this sample. To confirm this characterization, an additional TIE was conducted with powdered coconut charcoal (PCC). A 10% addition of PCC completely removed toxicity of 10% UN sediment (Table B-63). SIR-300 was not eluted because concentrations of metals in the sediment were all well below sediment quality guideline values, and the concentration of copper was well below the solid-phase LC50 for *E. estuarius*.

Ambersorb that was exposed to 10% sediment in the second solid-phase experiment was eluted for further Phase II HPLC fractionation that resulted in 30, one-mL methanol fractions. A portion of these fractions (300 µL) was reconstituted in clean dilution water, and exposures with *E. estuarius* were conducted. Complete mortality was observed in fractions 27 and 29. These fractions were analyzed for organochlorines, PCBs, and organophosphate and pyrethroid pesticides.

Solid-Phase Chemistry

The following measurements of chemicals in UN sediment are based on analysis of 100% sediment. Concentrations of chemicals measured in the undiluted sample were divided by ten to estimate the concentration of chemicals present in the sample used in the second solid-phase TIE (10%) (Table B-67).

Bifenthrin was detected in undiluted sediment at a concentration exceeding the sediment LC50s for *E. estuarius* (MPSL unpublished data). There were approximately 3.5 TU of bifenthrin in the undiluted sediment and an estimated 0.35 TU in 10% sample. The concentration of bifenthrin was also greater than the organic carbon-corrected LC50 in undiluted sediment. There was also approximately 0.05 TU permethrin in the undiluted sediment.

The concentrations of trace metals in the sediment were low. All metal concentrations were below their respective ERM values, except for silver in the undiluted sediment, and copper was well below the sediment LC50 for *E. estuarius*. The concentrations of organochlorines, PAHs, and PCBs were well below organism thresholds. Organophosphates were not detected in UN sediment.

Ambersorb from the second solid-phase TIE was fractionated and significant toxicity was observed in fractions 27 and 29. Analysis of these fractions detected several pyrethroids (Table B-67). Although there are no water-only LC50s for *E. estuarius*, concentrations of bifenthrin and permethrin in the fractions exceeded the water-only LC50s for *H. azteca*, a species that demonstrates comparable sensitivity to pyrethroids. Analysis of the solvent fractions for pyrethroids utilized the previously described method of direct injection into the gas chromatograph. It is possible that other compounds could be detected using this method, but these analyses were not performed. It is suspected that dilution occurred in the liquid-liquid solvent extraction employed prior to the original GC-MS analysis.

Interstitial Water TIEs

The initial test with IW caused complete mortality down to 10% sample (Table B-62). Three IW TIEs were conducted on a dilution series ranging from 10 to 100% UN sample. The 100% baseline sample in the first TIE contained 6.3 TU. Filtration reduced the toxicity to 1.2 TU (Table B-64). As with the CS IW sample, the UN sample contained few particles, so it is assumed that the contaminants sorbed to the filter material. Subsequent post-filter solid-phase extraction column treatments did not reduce toxicity further. Eluate from the sequential column treatment returned minimal toxicity to clean dilution water. Carboxylesterase enzyme added to IW significantly reduced toxicity and PBO added to IW significantly increased toxicity at all IW dilutions. The IW used in these treatments was not pre-filtered.

In the second IW TIE, additional centrifugation was used instead of filtration to remove particles from the sample. Centrifugation removed over half of the toxicity. Sufficient toxicity remained after centrifugation to apply other TIE treatments. Sample toxicity was not further reduced by either the cation or HLB columns (Table B-65). The IW sample was treated with sequential extraction by the cation column followed by the HLB column, and some toxicity was observed in the HLB eluate, after this sequential column treatment. Addition of the carboxylesterase enzyme again significantly reduced toxicity of this sample, and addition of PBO significantly increased toxicity.

In the third IW TIE, changes in toxicity with addition of the carboxylesterase enzyme and PBO (Table B-66) were emphasized. Bovine serum albumin (BSA – see Appendix A) was also added to account for possible reduced contaminant bioavailability through sorption to the protein-rich carboxylesterase enzyme. An additional centrifugation treatment was not used for several reasons. First, after the initial isolation of interstitial water by centrifugation, there were no particles visible in the sample. In addition, the previous TIE evidence suggested that toxicity was caused by pyrethroid pesticides, and there was concern about loss of these very hydrophobic chemicals through their sorption to the centrifuge tubes. Results of this TIE showed the HLB column significantly reduced toxicity, but the HLB methanol eluate did not return toxicity to clean dilution water. Enzyme addition also removed toxicity. The BSA blank did not remove toxicity and the PBO increased toxicity. These results suggest toxicity was caused by a pyrethroid pesticide.

Interstitial Water Chemistry

The concentrations of trace metals in the 100% IW were below known effects thresholds for *E. estuarius* (Table B-67). A low concentration of p,p'-DDE was measured, but total DDT was well below the published LC50 value. Low concentrations of PAHs were detected, but pyrethroids, PCBs, and organophosphates were not detected.

Weight of Evidence

Additions of SIR-300 combined with Ambersorb and PCC significantly reduced the toxicity of 10% UN sediment. These Phase I treatments characterized the cause of toxicity as a combination of a metal and an organic. Only the elution of the Ambersorb amendment returned toxicity, further characterizing the cause of toxicity as an organic. The addition of the carboxylesterase enzyme reduced toxicity, indicating the presence of a pyrethroid pesticide, but because a BSA blank was not conducted concurrently to differentiate between pyrethroid sorption and hydrolysis of the ester bond, the Phase II results were inconclusive. The presence of a toxic concentration of bifenthrin in the sediment, and the presence of toxic concentrations of several pyrethroids in the fractionated Ambersorb eluate confirms that pyrethroids contributed significantly to sediment toxicity.

The Phase I IW TIE treatments also characterized the cause of toxicity as an organic. Toxicity was reduced by both SPE columns, but only the HLB column returned toxicity. IW toxicity was also reduced by the addition of carboxylesterase enzyme, and increased by addition of PBO, indicating toxicity due to pyrethroid pesticides. The enzyme hydrolyzes ester-containing compounds, such as pyrethroid pesticides to their corresponding acid and alcohol, which are generally not toxic (Wheelock et al., 2004, 2006), and the PBO acts as a synergist in the presence of pyrethroids. There was no reduction of toxicity by the BSA blank, further indicating pyrethroids were contributing to toxicity. Although the Phase I and II TIE treatments indicated toxicity was caused by pyrethroid pesticides, analysis of the IW revealed only low concentrations of trace metals, organochlorines, and PAHs. Pyrethroids were not detected in the IW, but may have been present at concentrations below the detection limit, or lost in the TIE and analytical procedural steps discussed above.

3.3.4 San Diego Creek

As discussed above, the San Diego Creek station (SD) is a freshwater site approximately 0.5 km upstream of the Upper Newport Bay site (UN). A brief description of this watershed is provided above. Because the TIE and chemistry evidence from station UN suggested that pyrethroid pesticides were a possible cause of sediment toxicity to *E. estuarius*, TIEs on station SD also included treatments to address toxicity due to pyrethroids.

Solid-Phase TIEs

In the first solid-phase TIE, the addition of Ambersorb significantly reduced toxicity, but toxicity was also reduced in the dilution blanks, indicating the reduction of toxicity was caused by dilution, or through contaminant sorption to the sand used to dilute the sample (Table B-69). For unknown reasons, complete mortality was observed in the combinations of SIR-300 and Ambersorb. The carboxylesterase enzyme significantly reduced toxicity and the PBO significantly increased toxicity. A BSA blank treatment was not included in the first solid phase TIE. Because Type I pyrethroids have been demonstrated to be more toxic at lower temperatures, toxicity of SD sediment was assessed at 15 and 23°C. The 15°C baseline test was more toxic than the 23°C baseline. The addition of carboxylesterase also reduced toxicity in the 15°C test, but to a lesser extent than in 23°C. A second solid-phase TIE included additions of both the carboxylesterase enzyme, and BSA to account for sorption to the protein matrix of the enzyme. The enzyme significantly reduced toxicity, but there was also a reduction of toxicity with the addition of BSA. This result makes it difficult to differentiate between sorption to the enzyme and ester bond hydrolysis.

Elution of the Ambersorb produced significant toxicity when added back to clean dilution water (Table B-69). The Ambersorb solvent was fractionated using HPLC, and the thirty fractions were reconstituted and tested. Significant toxicity was observed in fractions 26 through 29. These fractions were analyzed individually for organochlorines, PCBs, organophosphates, and pyrethroids.

Solid-Phase Chemistry

The following measurements of chemicals in SD sediment are based on analysis of 100% sediment, which was the concentration that underwent the solid-phase TIEs (Table B-70). The concentrations of trace metals, organochlorines, PAHs, and PCBs were below the effects thresholds for *H. azteca*. Pyrethroids and organophosphate pesticides were not detected in SD sediment.

Ambersorb from the second solid-phase TIE was fractionated and significant toxicity was observed in fractions 26 through 29. Organophosphate analysis of fraction 26 revealed a significant concentration of dioxathion (94 µg/L). This chemical was not detected in the sediment because it was not on the sediment organophosphate analyte list. There is no published effect threshold for *H. azteca*, but the concentration in fraction 26 was greater than the LC50 for a similar amphipod, *Gammarus fasciatus* (LC50 = 8.6 µg/L; Mayer and Ellersiack, 1986). Toxic concentrations of bifenthrin and permethrin were also detected in several fractions.

Weight of Evidence

Phase I solid-phase TIE treatments characterized the cause of toxicity as organic. Additional Phase II treatments designed to identify pyrethroid pesticides were less conclusive. Although the enzyme reduced toxicity and the PBO and low temperature treatments increased toxicity, the reduction of toxicity with the addition of BSA confuses the conclusion that the cause of toxicity was a pyrethroid. The lack of pyrethroid detections in the sediment also confounds the results, but detection of pyrethroids in the toxic Ambersorb eluate fractions indicate that pyrethroids were playing a role in the toxicity of the sediment. As discussed above, detection of pyrethroids in the eluates without detection in the sediments could be attributed to insufficient sediment detection limits. It is also possible that the Ambersorb concentrated low levels of sediment pyrethroids during the exposure. Enzyme reduction as well as cold temperature and PBO potentiation can also indicate organochlorine toxicity, but most of these compounds were also below detection limits.

Dioxathion was also detected in the toxic fractions of the Ambersorb eluate. This chemical was a restricted use product that was used as an ingredient in miticides. It has not been registered for use as a product in the U.S. since the early 1990s.

3.3.5 Indiana Harbor

The Grand Calumet River and Indiana Harbor have been the subject of a number of sediment quality assessment studies. This area of northwestern Indiana is one of the most highly industrialized areas of the United States and is influenced by numerous point and non-point source discharges (Ingersoll et al., 2002; MacDonald et al., 2002). Sediments in this area are contaminated by complex mixtures of contaminants including metals, PAHs, PCBs, and organochlorine pesticides. MacDonald et al. (2002) reported that this area was among the most contaminated and toxic sites ever reported. These authors determined that metals, PAHs, and PCBs in sediments from this area were sufficient to account for observed toxicity to amphipods (*H. azteca*) in laboratory tests.

Solid-Phase TIEs

Initial tests with Indiana Harbor (IH) samples demonstrated complete mortality in undiluted sediment (Table B-71). The sediment test was repeated with a dilution series and significant mortality was observed in the 25 and 50% solid-phase concentrations.

Baseline survival in the first solid-phase TIE with 25% IH sediment was 36% (Table B-72). Ammonia concentrations in the overlying water were low, and unionized concentrations were well below the *H. azteca* LC50. It was assumed that IW unionized ammonia concentrations were higher, so a zeolite amendment was included in the TIE treatments. Although interstitial water ammonia concentrations were not measured, this treatment should have reduced interstitial water ammonia based on previous results with zeolite. The sample remained toxic despite zeolite addition, indicating ammonia probably contributed to toxicity, but was not the sole cause. Separate additions of SIR-300 and Ambersorb did not reduce toxicity, and a combination treatment comprised of all three sediment amendments also did not reduce toxicity. Addition of enzyme at both temperatures reduced toxicity, but a BSA blank was not included in the first solid-phase TIE. A second TIE was performed to account for any reduction of toxicity by sorption to the enzyme. Enzyme also reduced toxicity in this TIE, but a similar reduction was observed in the BSA blank indicating toxicity reduction could have been due to physical binding of contaminants.

Ambersorb addition did not reduce toxicity in the first solid-phase TIE, but toxicity was observed in the Ambersorb eluate. This suggests the cause of toxicity was partly due to an organic chemical. The second solid-phase TIE included a PCC treatment. Ten percent PCC addition also slightly reduced toxicity, further indicating toxicity caused by an organic chemical.

Solid-Phase Chemistry

The following measurements of chemicals in IH sediment are based on analysis of 100% sediment. Concentrations of chemicals measured in the undiluted sample were divided by four to estimate the concentration of chemicals present in the sample used in the solid-phase TIEs (25%, Table B-76).

The concentrations of several metals in the undiluted sediment were above the PEC sediment quality guidelines, including copper, which was also greater than the sediment LC50 for *H. azteca* (MPLS unpublished data). After dilution, the concentration of copper was below the sediment LC50, but sediment concentrations of several metals were above sediment quality guidelines.

Anthracene and total PAHs exceeded the PEC sediment quality guideline in both undiluted and diluted sediment. The sum of 13 PAH parent compounds in the diluted sediment, corrected for organic carbon, was approximately half the consensus-based median effect concentration of 1800 µg/g OC (Swartz 1999). PAH compounds were also summed using the target lipid model (Di Toro et al., 2000; Di Toro and McGrath, 2000). This model sums parent and transformation compounds and calculates species-specific toxic units for each compound. Based on this model, the concentrations of PAHs in undiluted sediment contributed approximately 0.72 TU, sufficient to contribute to toxicity of the 100% IH sediment. Toxic units of PAHs in the 25% IH sediment were estimated to be approximately 0.18 TU, which is near the lower limit of toxicity to *H. azteca* using the target lipid model (Joy McGrath, HydroQual, Inc., Mahwah, NJ, U.S., personal communication).

The concentration of total PCBs in 100% IH sediment exceeded the consensus-based extreme effect concentration of 1700 ng/g (MacDonald et al., 2000a). The concentration of chlordane in undiluted sediment exceeded the PEC, but was only half the PEC in diluted sediment. The concentration of organic carbon-normalized total DDT was well below published LC50 value (Nebeker et al., 1989). Pyrethroids and organophosphates were not detected in IH sediment.

Although the concentrations of PAHs in the undiluted sediment were substantial, the concentrations of PAHs in the Ambersorb eluate were low. No other chemicals were detected in the Ambersorb eluate.

Interstitial Water TIEs

Complete mortality was observed in all initial IW concentrations greater than 10% (Table B-71). The total IW ammonia concentration was 399 mg/L. Based on pH, temperature and salinity, the unionized ammonia concentration was 7 mg/L, greater than the unionized ammonia LC50 for *H. azteca* (5.44 mg/L). Three interstitial water TIEs were performed (Tables B-73 through B-75). The first TIE was designed to remove toxicity caused by ammonia. Complete removal of toxicity by these treatments would suggest ammonia was the sole cause of IW toxicity. The concentration of ammonia measured in the 20% baseline test exceeded the ammonia LC50 for *H. azteca*. Both zeolite column and air stripping reduced the ammonia below toxic levels, but toxicity was still observed (Table B-73). In addition, 5% IW was significantly toxic despite the fact that less than one-third TU of unionized ammonia was measured at this IW concentration. During the first few days of the IW TIE, the air stripping treatment reduced toxicity, but as the test progressed, blank toxicity was observed using this method (Table B-73).

After the IW sample had been treated with zeolite, TIE treatments designed to characterize toxicity due to metals and organics were conducted in the second IW experiment. Total ammonia in the IW was reduced to less than 1 mg/L compared to 2.9 TUs in the baseline sample, demonstrating the removal of a large portion of the toxicity by zeolite (Table B-74). Toxicity was further removed by the cation column, the HLB column, and the sequential columns treatment. Toxicity was not returned in any of the column eluates. The enzyme did not significantly reduce toxicity, and the BSA blank demonstrated that the small reduction was probably due to sorption. PBO addition increased toxicity to 7.5 TU.

The third interstitial water TIE emphasized elution of toxic concentrations of chemicals from the HLB and cation columns. Acetone and methanol were used separately to elute two HLB columns. For this TIE, the sample was passed separately through the cation and HLB columns before being passed through a zeolite column. Baseline toxicity in this TIE was greater than baseline toxicity in the second TIE (Table B-75). The cation column removed a small amount of toxicity, but the cation eluate did not return any toxicity, suggesting cationic metals were probably not a primary cause of toxicity. The HLB column significantly removed toxicity, and when acetone and methanol eluates were spiked separately into dilution water, significant toxicity was observed in both cases. The methanol eluate was analyzed for chemistry.

Interstitial Water Chemistry

Both undiluted and diluted IW contained concentrations of copper, lead and zinc that were greater than the LC50s for *H. azteca* (Table B-76), but the TIE results suggest that cationic metals were not the primary cause of toxicity. In the second IW TIE, the sample was passed through a zeolite column before being subjected to the other treatments and this could have reduced the concentration of metals in the IW (Burgess et al., 2004). The IW sample in the third TIE was passed through the zeolite column after the other column treatments to avoid loss to zeolite. It was expected that toxic concentrations of metals would be extracted with the cation column and returned in the cation eluate, but toxicity reduction was not observed, and the eluate only returned a small amount of toxicity. The IW copper concentration was 262 μg/L. Based on the results of the copper spiking experiments, the cation column would have extracted this concentration of copper and returned it in the cation eluate. The fact that the cation eluate did not return toxicity indicted that the metals were not bioavailable.

The concentration of fluoranthene and anthracene in undiluted and diluted IW were also greater than their respective LC50s for *H. azteca* (44.9 μg/L, Suedel et al., 1993; 5.6 μg/L, Hatch and Burton, 1999). Although dissolved organic carbon was not measured in the interstitial water, the target lipid model was employed using an estimated DOC of 50 mg/L (Williamson and Burgess, 2003). The results of the target lipid model indicate there were approximately 7.6 TU of PAHs in diluted interstitial water. PAHs were also the only compounds detected in the eluate from the HLB column, but based on the target lipid model, these compounds were not present at toxic concentrations.

The concentration of total PCBs was converted to an aroclor concentration (NOAA, 1989) and compared to a published lowest observed effect concentration (LOEC). Based on this LOEC, there was approximately one-third of a toxic unit of total PCB in undiluted IW. Low concentrations of p,p'-DDE and endosulfan I were also measured. Pyrethroids and organophosphates were not detected in UN IW.

Weight of Evidence

Of the solid-phase Phase I TIE methods used to treat 25% IH sediment, only the addition of carboxylesterase enzyme, BSA and PCC reduced toxicity. Reduction of toxicity with these treatments characterized the cause of toxicity as organic. The Phase II elution of the Ambersorb amendment returned toxicity, further characterizing the cause of toxicity as an organic, but only low concentrations of PAHs were detected in the eluate. Indiana Harbor sediment was the most contaminated ambient sample in this study. The sediment contained high concentrations of ammonia, and concentrations of metals, PAHs and PCBs exceeded sediment quality guidelines and published tolerance thresholds. Although some metals exceeded guideline values, the TIE data suggest that metals did not play a significant role as the cause of toxicity.

The Phase I IW TIE treatments also characterized the cause of toxicity due to a combination of unionized ammonia and organic chemicals. Toxicity was reduced by both SPE columns, but the cation column did not return toxicity and there was no reduction of toxicity with the addition of EDTA, suggesting the metals were not bioavailable. Although IW toxicity was reduced by the addition of carboxylesterase enzyme, and increased by addition of PBO, the reduction of toxicity by the BSA addition indicates that toxicity was caused by an organic chemical other than a pyrethroid pesticide. The Phase II elution and analysis of the HLB eluate detected PAHs, and IW concentrations of fluoranthene and anthracene were higher than *H. azteca* LC50s. The weight of evidence suggests that toxicity was caused by ammonia and selected PAHs, but as already discussed, this conclusion is tempered by uncertainties over loss of hydrophobic contaminants during the elution processes and during sample preparation prior to chemical analysis.

3.3.6 Alisal Creek

Alisal Creek (AC) is a freshwater site located near the southeastern edge of the City of Salinas, in central California. The Salinas Valley is among the most intensively farmed regions in California and year-round agriculture in this area supply as much as 80% of the lettuce consumed in the United States. Alisal Creek is part of the Gabilan Creek watershed and flows to this part of system are dominated by agriculture return water from the lower Salinas Valley, combined with some mixed-use urban runoff from the City of Salinas. This site has been monitored by the Central Coast Regional Water Quality Control Board for a number of years, and recent studies have indicated ambient water toxicity associated with organophosphate (OP) pesticides and sediment toxicity associated with OP, organochlorine and pyrethroid pesticides (Hunt et al., 2006).

Solid-Phase TIE

Sediment from AC was significantly toxic in initial tests and the solid-phase TIE was conducted on undiluted sample (Table B-77). In the solid-phase TIE, Ambersorb and SIR-300 sediment amendments did not reduce toxicity either individually or in combination (Table B-78). Toxicity was reduced by addition of the carboxylesterase enzyme to sediment overlying water. Addition of BSA did not reduce toxicity, indicating toxicity was due to a pyrethroid pesticide rather than sorption to the enzyme. Effects of temperature were inconclusive because there was complete mortality in the baseline at both 15 °C and 23 °C. The effects of PBO were also inconclusive because of complete mortality in the baseline. Although neither the SIR-300 nor

Ambersorb amendments reduced toxicity, both were eluted to determine if toxic concentrations of chemicals sorbed to either resin. The Ambersorb elution returned significant toxicity to clean dilution water (0% survival), but the SIR-300 did not. Some blank toxicity was observed in acetone eluate from the Ambersorb added to the control sediment (47% survival).

Sediment Chemistry

The concentration of lambda-cyhalothrin in 100% AC sediment was greater than the total and organic carbon corrected sediment LC50s for *H. azteca* (Amweg et al., 2005). (Es)fenvalerate was also detected in this sample (Table B-81). Several organochlorine pesticides were detected at concentrations greater than sediment quality guideline values. The sediment contained approximately one-half TU of organic carbon-normalized DDT. The concentrations of metals, PAHs, and PCBs were low relative to guideline values and published toxicity thresholds, and there were no organophosphates detected in AC sediment.

Significantly toxic concentrations of bifenthrin, cypermethrin, and permethrin were detected in the Ambersorb eluate. Although there were no organophosphates detected in the sediment, there was a significant concentration of chlorpyrifos in the Ambersorb eluate (~ 4 TU). Diazinon was also detected in the eluate, as were low concentrations of o,p'-DDD and p,p'-DDE. Detection of organophosphates in the eluates without detection in the sediments could be attributed to insufficient sediment detection limits and the possibility that the Ambersorb concentrated low levels of sediment organophosphates during the exposure.

Interstitial Water TIEs

The initial IW test demonstrated significant toxicity down to 10% sample. Two IW TIEs were conducted with 50% AC IW. Baseline toxicity in the first TIE contained 7.7 TU (Table B-79). Toxicity was reduced by the cation column, the HLB column and the sequential column treatments. The eluate from the individual HLB column was the only treatment that returned toxicity. The carboxylesterase enzyme also significantly reduced IW toxicity. Addition of BSA did not reduce toxicity while addition of PBO increased toxicity. These results indicate toxicity due to a pyrethroid pesticide. A second IW TIE was conducted which emphasized HLB column elution. Results were obtained similar to those described above (Table B-80), and the HLB eluate was analyzed for organic contaminants.

Interstitial Water Chemistry

AC IW contained concentrations of copper, lead, and zinc greater than their published LC50s for *H. azteca* (Table B-81). Solid-phase extraction with the cation column reduced toxicity, but toxicity was not observed in the cation eluate. Similarly, the sequential cation HLB column extraction reduced toxicity, but the eluate from the cation column did not return toxicity. It is possible that metals contributed to toxicity of this sediment, but the results suggest metals played a limited role, perhaps due to reduced bioavailability.

Permethrin was detected in the undiluted and diluted interstitial water at a concentration greater than the water-only LC50 (Anderson el al., 2006). Lambda-cyhalothrin also was detected. Concentrations of two DDT metabolites were greater than published LC50s for *H. azteca* (Hoke et al., 1994). The concentrations of PAHs were low, and PCBs and organophosphates were not detected.

Weight of Evidence

Toxicity of AC sediment was reduced by addition of the carboxylesterase enzyme to the overlying water. Reduction of toxicity by the enzyme, but not the BSA, is a Phase I TIE method to characterize toxicity due a pyrethroid pesticide. Return of toxicity with the elution of the Ambersorb further characterized the toxicity as organic, and Phase II analysis of the eluate revealed toxic concentrations of chlorpyrifos and several pyrethroids. Chemical analysis of AC sediment demonstrated 3.1 organic carbon-normalized toxic units of lambda-cyhalothrin.

In the Phase I IW TIE the cation and HLB columns reduced toxicity, characterizing the cause of toxicity as a combination of metals and organics. Using the Phase II SPE elution methods, only the HLB eluate returned toxicity. Results using the carboxylesterase enzyme, BSA, and PBO methods were useful in characterizing and identifying toxicity due to a pyrethroid pesticide. Although there were 3 TUs of metals in the IW, no toxicity was observed in the cation column eluate, and this suggests metals probably played a minimal role in causing toxicity. There were 14.3 TU of permethrin in the IW, but there were no pyrethroids detected in the HLB eluate.

The chemical analysis, combined with the toxicity results, identified several chemicals that were above the tolerance threshold for *H. azteca*, but the concentrations of lambda-cyhalothrin in the solid-phase and permethrin in the interstitial water demonstrate pyrethroids were probably the primary cause of toxicity. The presence of toxic concentrations of pyrethroids and chlorpyrifos in the Ambersorb eluate suggests these pesticides also contributed to toxicity of this sample. As with the TIEs described above, the methods for Ambersorb and HLB column elution need refinement to ensure complete elution of hydrophobic chemicals from the extraction media. In addition, the research group suspects the lack of detection of selected chemicals in these eluates was also due to loss and dilution in the sample preparation steps prior to gas chromatography.

Table 3-1. Results Summary and Timeline for TIEs Conducted on Sediment from Switzer Creek.

Switzer Creek			
Solid-Phase	**Procedure**	**Result**	**Interpretation**
11/23/04	Sediment TIE (25% Sample)	Ambersorb and SIR-300 reduce toxicity	Toxicity caused by combination of metals and organics
12/17/04	Elute Ambersorb and SIR-300	Ambersorb eluate toxic, SIR-300 eluate not toxic	Toxicity caused by organics
2/11/05	Repeat Sediment TIE (25% Sample)	Ambersorb and SIR-300 reduce toxicity	Toxicity caused by combination of metals and organics
2/28/05	Repeat Elute Ambersorb and SIR-300	Ambersorb eluate toxic, SIR-300 eluate not toxic	Toxicity caused by organics
5/20/05	Fractionate Ambersorb Eluate with HPLC	Fraction 27 toxic	Contaminant is very non-polar organic
10/10/05	Tissue Loading TIE (10% Sample)	Ambersorb reduces toxicity	Toxicity caused by organics
1/18/06	Fraction 27 Analyzed for OC, PAH, Pyrethroids	Trans chlordane detected	Chlordane below NOEC
1/18/06	Tissue analyzed for metals, OC, PCB, PAH, Pyrethroids	Oxadiazon was only chemical detected in baseline tissue. No chemicals detected in treatment tissues.	
	Sediment chemical analysis	See Tables 9 and B55	

Table 3-2. Results Summary and Timeline for TIEs Conducted on Sediment and Interstitial Water from Consolidated Slip.

Consolidated Slip			
Solid-Phase	**Procedure**	**Result**	**Interpretation**
6/24/05	Sediment TIE (1% Sample)	Ambersorb and SIR-300 reduce toxicity	Toxicity caused by combination of metals and organics
8/5/05	Elute Ambersorb and SIR-300	Ambersorb eluate toxic, SIR-300 eluate not toxic	Toxicity caused by organics
10/10/05	Ambersorb Acetone Eluate HPLC Fractionation	No HPLC fractions were toxic	Contaminant is too non-polar to be removed from HPLC column with methanol
11/7/05	Ambersorb Eluate analyzed for OC, OP, PAH, PCB, Pyrethroids	Small amounts of PAHs detected	
Interstitial Water	**Procedure**	**Result**	**Interpretation**
6/24/05	Interstitial Water TIE	Filtration reduces toxicity	No particulates in sample therefore contaminants bound to filter
7/29/05	Interstitial Water TIE (centrifugation substitute)	Cation column and HLB column reduce toxicity	Toxicity caused by organics
		Only HLB eluate returns toxicity	
8/19/05	Interstitial Water TIE with serial HLB elution	Only 100% methanol elution toxic	Contaminant is very non-polar
9/8/05	Analyze HLB methanol fraction	Small amounts of PAHs detected	
	Sediment and interstitial water chemical analysis	See Tables 9 and B61	

Table 3-3. Results Summary and Timeline for TIEs Conducted on Sediment Interstitial Water from Upper Newport Bay.

Upper Newport Bay			
Solid-Phase	**Procedure**	**Result**	**Interpretation**
6/10/05	Sediment TIE (100% Sample)	No treatments reduce toxicity	Must dilute sediment
7/22/05	Sediment Dilution Test	LC50 = 5.9% sediment	10% sample optimal for TIE
8/12/05	Sediment TIE (10% Sample)	Ambersorb and SIR-300 reduce toxicity when added in combination	Toxicity caused by combination of metals and organics
9/26/05	Elute Ambersorb	Ambersorb eluate toxic	Toxicity caused by organics
9/29/05	Fractionate Ambersorb Eluate with HPLC	Fractions 27 and 29 toxic	Contaminant is very non-polar organic
12/2/05	Sediment TIE with PCC	PCC reduces toxicity	Supports toxicity caused by organics
1/18/06	Analyze fractions 27 and 29 for OC, PCB, OP, Pyrethroids	No detectable chemicals in fractions 27 and 29	
Interstitial Water	**Procedure**	**Result**	**Interpretation**
6/10/05	Interstitial Water TIE	Filtration reduces toxicity	No particulates in sample therefore contaminants bound to filter
		Some add-back in HLB eluate	Some organic contaminants present
		Carboxylesterase Enzyme reduced toxicity and PBO increased toxicity	Pyrethroid signature
8/5/05	Interstitial Water TIE (centrifugation substitute)	All columns reduce toxicity but only return is with HLB methanol eluate	Toxicity caused by organics
		Enzyme reduced toxicity and PBO increased toxicity	Pyrethroid signal
10/21/05	Interstitial Water TIE (centrifugation substitute)	Reduction of toxicity with HLB but no add-back with eluate	Contaminant is very non-polar
		Reduction of toxicity with enzyme, and increase of toxicity with PBO and BSA	Pyrethroid signal
	Sediment and interstitial water chemical analysis	See Tables 9 and B67	

Table 3-4. Results Summary and Timeline for TIEs Conducted on Sediment from San Diego Creek.

San Diego Creek			
Solid-Phase	**Procedure**	**Result**	**Interpretation**
6/17/05	Sediment TIE (100% Sample)	Ambersorb reduces toxicity	Toxicity caused by organics
		Enzyme reduces toxicity and PBO increases toxicity	Pyrethroid signature
		15 degree treatment increases toxicity	Pyrethroid signature
7/8/05	Eluate Ambersorb	Ambersorb eluate toxic	Toxicity caused by organics
10/10/05	Ambersorb Acetone Eluate HPLC Fractionation	HPLC Fractions 26-29 toxic	Contaminant is very non-polar
11/18/05	Sediment TIE with enzyme and BSA treatments	Both treatments reduce toxicity	Toxicity not caused by pyrethroids
1/18/06	Analyze Fractions 26-29 for OC, PCB, OP, Pyrethroids	Toxic concentration of dioxathion detected in fraction 26	
	Sediment and interstitial water chemical analysis	See Tables 9 and B70	

Table 3-5. Results Summary and Timeline for TIEs Conducted on Sediment Interstitial Water from Indiana Harbor.

Indiana Harbor			
Solid-Phase	**Procedure**	**Result**	**Interpretation**
8/12/05	Sediment TIE (25% Sample)	Enzyme reduced toxicity, Ambersorb did not reduce toxicity.	Organics bound by enzyme, or pyrethroids reduced by enzyme
11/18/05	Sediment TIE with enzyme, BSA, and PCC	PCC reduces toxicity. Enzyme and BSA also reduce toxicity.	Toxicity caused by organics, but not by pyrethroids
12/2/05	Elute Ambersorb	Ambersorb eluate toxic	Toxicity caused by organics
1/18/06	Analyze Ambersorb eluate for OC, PCB, OP, PAH, Pyrethroids	Low concentrations of PAHs detected in eluate	
Interstitial Water	**Procedure**	**Result**	**Interpretation**
8/2/05	Interstitial Water TIE with ammonia treatments	Zeolite reduces toxicity but does not eliminate it	Ammonia is a primary cause of toxicity, but other contaminants present
9/30/05	Interstitial Water TIE (after passing through zeolite columns)	All columns reduce toxicity but no add back in any eluates	Contaminant bound to or in columns. ELISA shows 2 TU chlorpyrifos
9/30/05	Interstitial Water TIE (then passed through zeolite columns)	HLB column reduced toxicity. Eluates returned toxicity	Toxicity caused by organics
1/18/06	Analyze HLB eluate for OC, PCB, OP, PAH, Pyrethroids	Low concentrations of PAHs detected in eluate	
	Sediment and interstitial water chemical analysis	See Tables 9 and B76	

Table 3-6. Results Summary and Timeline for TIEs Conducted on Sediment Interstitial Water from Alisal Creek.

Alisal Creek			
Solid-Phase	**Procedure**	**Result**	**Interpretation**
8/19/05	Sediment TIE (100% Sample)	Enzyme reduced toxicity, BSA did not reduce toxicity	Pyrethroids present
10/18/05	Elute Ambersorb	Ambersorb eluate toxic	Toxicity caused by organics
12/2/05	Elute SIR-300	SIR-300 eluate not toxic	Toxicity not caused by metals
1/18/06	Analyze Ambersorb eluate for OC, PCB, OP, PAH, Pyrethroids	Low concentrations of organochlorines and a toxic concentration of chlorpyrifos detected	
Interstitial Water	**Procedure**	**Result**	**Interpretation**
8/26/05	Interstitial Water TIE	All columns reduce toxicity, but only return is with HLB eluate	Some organic contaminants present
		Enzyme reduced toxicity and PBO increased toxicity	Pyrethroids present
10/28/05	Interstitial Water TIE	HLB column reduces toxicity, and eluate returns toxicity	Toxicity caused by organics
1/18/06	Analyze HLB eluate for OC, PCB, OP, PAH, Pyrethroids	Low concentrations of organochlorines detected	
	Sediment and interstitial water chemical analysis	See Tables 9 and B81	

Table 3-7. Summary Scorecard of TIE Results for Marine Ambient Samples.

		Phase I – Characterize – Alter toxicity through TIE process	Phase II – Identify – Separate toxic components from non-toxic, test, and analyze	Phase III – Confirmation – Evaluate chemistry with species sensitivity	Probable Cause
Switzer Creek	**Solid-Phase**	Ambersorb and SIR-300 reduced toxicity – metals and organics	Elution of Ambersorb returns toxicity – organics – toxic fraction analyzed - chlordane detected in eluate, but well below threshold	Concentration of pyrethroids above organism thresholds –pyrethroids were not detected in eluate	Pyrethroids
Consolidated Slip	**Solid-Phase**	Ambersorb and SIR-300 reduced toxicity – metals and organics	Elution of Ambersorb returns toxicity – organics – PAHs detected in eluate	Total PAH in sediment exceeded target lipid model threshold and *R. abronius* LC50 (Page et al., 2002) – eluted PAHs did not exceed water-only thresholds	Total PAH
	Interstitial Water	HLB column and cation column reduced toxicity – metals and organics	Elution of HLB column returns toxicity – organics – PAHs detected in eluate	PAHs did not exceed water-only thresholds	Total PAH
Upper Newport	**Solid-Phase**	Ambersorb and SIR-300 reduced toxicity – metals and organics	Elution of Ambersorb returns toxicity – organics – no chemicals detected in eluate	Concentration of bifenthrin above organism threshold – pyrethroids not detected in eluate	Bifenthrin
	Interstitial Water	HLB column and cation column reduced toxicity – metals and organics	Elution of HLB column returns toxicity – organics – no chemicals detected in eluate	Only low concentrations of metals, organochlorines and PAHs detected	Organic

Table 3-7 Continued. Summary Scorecard of TIE Results for Freshwater Ambient Samples.

		Phase I – Characterize – Alter toxicity through TIE process	Phase II – Identify – Separate toxic components from non-toxic, test, and analyze	Phase III – Confirmation – Evaluate chemistry with species sensitivity	Probable Cause
San Diego Creek	**Solid-Phase**	Ambersorb reduced toxicity - organics	Elution of Ambersorb returns toxicity – organics – toxic fractions analyzed – dioxathion detected in one of the toxic fractions	Concentration of dioxathion greater than LC50 for similar test organism	Dioxathion
Indiana Harbor	**Solid-Phase**	Addition of PCC, enzyme, and BSA reduce toxicity - organics	Elution of Ambersorb returns toxicity – organics –only PAHs detected	Concentrations of metals, PAHs, and PCBs greater than guidelines or thresholds	Ammonia, PAHs, PCBs
	Interstitial Water	Zeolite partially reduces toxicity – ammonia – enzyme and BSA reduce toxicity - organic	Elution of HLB column returns toxicity – organic – PAHs detected in eluate – no	Concentrations of ammonia, metals and PAHs greater than thresholds	Ammonia, PAHs
Alisal Creek	**Solid-Phase**	Enzyme reduced toxicity and BSA did not - pyrethroid	Elution of Ambersorb returns toxicity – organics – chlorpyrifos and low concentrations of DDT metabolites detected	3.1 TU (oc normalized) lambda-cyhalothrin in sediment and 4.4 TU chlorpyrifos in Ambersorb eluate	Lambda-cyhalothrin and chlorpyrifos
	Interstitial Water	HLB column and cation column reduced toxicity – metals and organics	Elution of HLB column returns toxicity – organics – no chemicals detected in eluate	3 TU metals and 14.3 TU permethrin	Metals, Permethrin

3.3.7 Method Evaluation Based on Ambient Sample TIE Results

In the final phase of this project, existing sediment TIE methods were used to investigate causes of solid-phase and interstitial water toxicity using three ambient marine and freshwater samples. In addition to standard methods, these TIEs also incorporated novel methods developed as part of this project. The results suggest that these methods were successful at differentiating between sediment toxicity caused by ammonia, metals, and organic chemicals in both marine and freshwater samples. In the majority of cases, treatment of solid-phase samples with either Ambersorb, SIR-300 or powdered coconut charcoal reduced toxicity. In the one sample containing toxic concentrations of unionized ammonia (Indiana Harbor), addition of zeolite was effective at reducing this constituent prior to application of additional Phase I treatments. Where interstitial water toxicity was observed in samples, Phase I treatments usually were effective at characterizing toxicity due to metals, organics or ammonia. The Phase I TIE methods were particularly useful when the results of solid-phase and interstitial water procedures were considered together. One caveat to this conclusion is that in almost all cases, the various TIE treatments were less effective in highly toxic samples. Almost all samples required some degree of dilution in order to demonstrate differences between TIE treatments and baseline toxicity. Possible effects of sample dilution on TIE interpretation are discussed in Section 3.4.5, and suggested research topics to improve solid-phase and interstitial water Phase I treatments are discussed in Section 3.5 (Future Research).

Mixed results were observed using Phase II TIE methods. As part of the Phase II TIE solid-phase procedures, the resin amendments (SIR-300 or Ambersorb) were recovered from the sediments and eluted with either acid (SIR-300) or acetone (Ambersorb) to determine whether toxic concentrations of cationic metals or organic chemicals were returned to clean water spiked with the eluates. Similar procedures were used to elute metals or organic chemicals from the cation and HLB columns used in the interstitial water TIEs. In all of the TIEs, significant toxicity was observed in water spiked with the acetone Ambersorb eluates. In three of the four interstitial water TIEs, toxicity also was observed in water spiked with the HLB methanol eluates. Toxicity was not observed in waters spiked with acid eluates from the SIR-300 or the cation columns. These results were useful to confirm that organic chemicals removed by the extraction media were eluted with solvents and spiked into clean water at toxic concentrations. However, the methods were largely unsuccessful at identifying the chemicals responsible for eluate toxicity in these samples, which is a key goal of the Phase II TIE process. In three of the TIEs, the Ambersorb eluates were subjected to HPLC methanol fractionation and the toxic fractions were analyzed using GC-MS. The remaining Ambersorb acetone eluates were analyzed without further fractionation as were all of the HLB column methanol eluates. In most cases, relatively low concentrations of organic chemicals were detected in both the HPLC methanol fractions, and in the Ambersorb and HLB eluates. The TIE evidence suggests that in at least three of these samples, toxicity likely was due to pyrethroid pesticides (e.g., Switzer Creek, Upper Newport Bay, and Alisal Creek). As discussed above, these pesticides appear to be particularly problematic because they are highly toxic to amphipods, and are very hydrophobic. These characteristics make these and other hydrophobic compounds difficult to elute, and subject to loss through sorption and dilution at different steps in the TIE and analytical chemistry processes. In Section 3.5 further research is proposed to address modifications to the current methods as they apply to Phase II elution procedures.

The results suggest that the carboxylesterase enzyme is a useful Phase II TIE treatment for identifying sediment toxicity due to pyrethroid pesticides. Addition of the enzyme to sediment overlying water or interstitial water reduced toxicity of Upper Newport Bay and Alisal Creek sediments, both of which contained pyrethroid concentrations above published LC50s (the enzyme was not used in the Switzer Creek TIE). In both cases, additions of Bovine Serum Albumin did not affect sample toxicity, and the reduction of toxicity could therefore be ascribed to pyrethroid hydrolysis by the enzyme rather than sorption. Addition of the metabolic inhibitor PBO increased toxicity of these samples, and this provided additional evidence of toxicity due to pyrethroid pesticides. In the Indiana Harbor TIE, a reduction of toxicity also was found with addition of enzyme and BSA, and an increase in toxicity with addition of PBO. In this case, the results make it difficult to differentiate between hydrolysis of pyrethroids and sorption of other organic chemicals to BSA. Although PBO increased toxicity of Indiana Harbor sediment, the absence of measurable pyrethroids suggests toxicity was not caused by these compounds, and that the synergism was due to other chemicals. The results indicate that the carboxylesterase enzyme provides a useful line-of evidence when used in conjunction with BSA, PBO, and chemical analyses.

The primary approach to Phase III (confirmation) TIEs in the current study was to compare chemical concentrations in sediments and interstitial waters to known effect levels derived from dose-response experiments (i.e., species sensitivity approach). When these data were not available, concentrations were compared to empirically-derived sediment quality guideline values, or model-based values such as the target lipid model for PAHs. In the six ambient sample TIEs presented above, the Phase I and II TIE evidence suggested that toxicity was caused by organic chemicals and ammonia. Concentrations of selected organic chemicals exceeded LC50 values in solid-phase samples from Switzer Creek, Upper Newport Bay, Alisal Creek and Indiana Harbor, and provided confirmation that these chemicals were present at toxic concentrations. The lack of toxic concentrations of these chemicals in the majority of Ambersorb or HLB eluates prevents a final evaluation of the utility of the species sensitivity approach in these TIEs. Toxic concentrations of copper were measured in interstitial water from Alisal Creek and Indiana Harbor, but the lack of toxicity in the cation column eluate in these TIEs suggests that copper cation concentrations did not exceed toxicity thresholds in these samples.

Many of these samples were highly contaminated by complex chemical mixtures. Sediments from Indiana Harbor and Consolidated Slip contained high concentrations of PAHs, but there are no LC50 values for most of these constituents, particularly toxicity of PAH transformation products, to the amphipods used in this evaluation. In this case, PAH concentrations were compared to guideline values or to model-derived toxicity values which consider parent compounds and their transformation products. This approach provided evidence that PAHs were contributing to the toxicity of these samples. Indiana Harbor sediment also contained high concentrations of metals, PCBs and other organic chemicals, and these were also compared to guideline and LC50 values. This provided additional lines-of-evidence, but illustrates the difficulty of confirming toxicity that is most likely due to the interaction of a number of organic constituents and unionized ammonia. These results suggest that Phase III TIE methods require additional refinement.

3.4 Confounding Factors

In addition to the effects of xenobiotic contaminants, there are a number of physical and natural chemical factors that may adversely affect organisms in sediment toxicity tests. In TIEs where none of the treatments listed in this report are effective at reducing toxicity, these factors should also be considered. These include sediment grain size, ammonia and hydrogen sulfide, ionic constituents, and in the case of freshwater tests specifically, hardness, alkalinity, and conductivity. In addition to factors that directly affect organisms, sample collection and storage procedures can affect chemical composition and bioavailability in field-collected samples, and thus influence test results. The following section describes some of the most common confounding influences in sediment toxicity tests and provides guidance on addressing them in the TIE process. Emphasis is placed on factors that may influence test results with amphipods, but this discussion is applicable to other sediment and interstitial water test species (see also Bay et al., 2003; Waller et al., 2005).

3.4.1 Ammonia and Hydrogen Sulfide

Unionized ammonia and hydrogen sulfide are common constituents, particularly in organically enriched anoxic sediments, and both have the potential to confound interpretation of toxicity test results. Detailed methods to address unionized ammonia toxicity are presented above. In the majority of cases, hydrogen sulfide concentrations are quickly attenuated during sample mixing, and through aeration of test chambers, so toxicity due to hydrogen sulfide may only occur in highly anoxic sediments. As with ammonia, hydrogen sulfide concentrations can be compared to known toxicity thresholds, when available, to determine if they are within toxic ranges (Bagarino 1992; Phillips et al., 1997). Knezovich et al. (1996) presented tolerance values for four marine species used in marine sediment and interstitial water toxicity testing. Because of its volatility, toxicity due to hydrogen sulfide may be reduced using aeration treatments. In addition, hydrogen sulfide toxicity is pH dependant, with a greater percentage of the toxic hydrogen sulfide species formed at lower pH. Increased toxicity at lower pH can be used to indicate toxicity due to hydrogen sulfide.

3.4.2 Grain Size

Sediment grain size may affect some amphipod species and should be considered when TIE treatments are ineffective at reducing toxicity. There is some evidence to suggest survival of *E. estuarius* is negatively correlated with high clay concentrations in field sediments (Hunt et al., 2001). DeWitt et al. (1989) provided evidence of mortality of this species in sediments with high percent fines, and provided a regression equation to account for mortality in fine grained sediments. Tay et al. (1998) found this genus can tolerate sediments with up to 70% clay. Hunt et al. (2001) found that mortality of *E. estuarius* was highly correlated with the clay concentration of San Francisco Estuary sediments. Because chemical concentrations co-varied with percent fines and percent clay in these sediments, it was not possible to determine whether high concentrations of clay caused amphipod mortality in this study. In situations where sediments contain high percentages of clay, and application of the TIE procedures described in this report do not remove toxicity, researchers are advised to consider grain size distributions as a possible source of amphipod mortality. Another marine amphipod species, *Rhepoxynius abronius* naturally occurs in sandy marine habitats and does not tolerate high silt or high clay sediments, presumably because they impair gill function. DeWitt et al. (1988) found that survival of *R. abronius* may decline in fine-grained sediments, and also developed regression procedures to account for grain size effects with this species. U.S. EPA (1994c) recommends not using *R. abronius* for assessing sediments having greater than 90% fines. In situations where sample grain size distributions are unknown prior to testing, U.S. EPA (1994c) recommends including silt/clay reference sediments having particle size distributions that bracket test sediments. There is no evidence to suggest sediment grain size has any appreciable effect on survival of *H. azteca*.

3.4.3 Effects of Ions and Other Chemical Constituents

Acceptable levels of conductivity and hardness are stipulated in most freshwater toxicity test protocols, and when these levels are exceeded, their influence on test results should be considered. In addition to these factors, a number of ions have been shown to influence test organism response when their concentrations deviate beyond tolerance limits. Mortality can result from depleted concentrations of essential ions, or when concentrations of specific ions exceed their respective toxicity thresholds (Waller et al., 2005). There are several investigative approaches to determine if ionic imbalance is contributing to toxicity. Goodfellow et al. (2000) summarize major ion toxicity and provide a number of TIE approaches. These include the use of computer models, synthetic waters, ion add-back, and restoration of ion balance (Waller et al., 2005). Some naturally occurring interstitial water chemicals may be compared to published toxicity thresholds. For example, Lasier et al. (1997) showed that bicarbonate concentrations above 600 mg/L (as $CaC0_3$) were toxic to *H. azteca,* and recommended that bicarbonate be considered in the list of routine measures because it is common, particularly in estuarine interstitial waters. Lasier et al. (2000) also showed that manganese is toxic to *H. azteca* at concentrations commonly found in interstitial water, and should be included in chemical analyses of freshwater sediments. Burgess et al. (2007) recently presented solid-phase TIE methods for the anionic metals arsenic and chromium, using anion exchange resins. Use of these resins should be considered in situations where the standard list of TIE methods listed in the current report is ineffective.

3.4.4 Sample Storage and Manipulation

Sediment toxicity testing guidance documents recommend that testing be initiated within 14 days of sample collection (ASTM 2000). This is not always practical in sediment TIEs for a number if reasons. Because TIEs are iterative, with each step building on results from previous steps, a large volume of sediment is required and considerable time may elapse between treatments. Large volumes of sediment are also required to produce sufficient interstitial water for all treatments. To reduce variability introduced by separate sediment samples, it is convenient to use a single sample for all TIE procedures and this may result in storing sediment for long periods as the TIEs progress. For example, the researchers collected as much as 40L of sediment for the ambient TIEs presented in this report, and these sediments were stored for as long as one year. Extended holding time can change the toxicity of stored sediment and interstitial water (Malueg et al., 1986; Robinson et al., 1988; Doe et al., 2003). Changes in toxicity over time can be quantified by comparing results of the baseline tests conducted with each TIE step. Of particular concern in organic-rich sediments is the increased production of ammonia and hydrogen sulfide during extended storage times (Doe et al., 2003). Water quality parameters such as pH, dissolved oxygen, and salinity or conductivity also must be closely monitored as they may contribute to toxicity and may change during extended storage (Waller et al., 2005).

3.4.5 Sediment Dilution Effects

Diluting the sediment for a solid-phase TIE can significantly affect the chemistry of the sample and therefore influence interpretation of the TIE results. The role of contaminants that are marginally toxic in an undiluted sample may be reduced and overlooked in TIEs conducted at lower sediment or interstitial water concentrations. The research group recommends conducting solid-phase and interstitial water TIE treatments on both diluted and undiluted sediment to address this concern. As discussed above, solvent elution of carbonaceous resins and acidic elution of cation resins are particularly informative using undiluted solid-phase samples, and data from these treatments can be combined with information from lower dilutions to add to the weight-of-evidence in a TIE. Interstitial water TIEs should always be conducted using a dilution series to allow for the calculation of toxic units (TUs).

To facilitate separation of resins used in solid-phase TIEs, the SZ sediment was pre-sieved to reduce the concentrations of coarse grain fractions. Pre-sieving may alter the chemical composition of samples by increasing the relative concentrations of contaminants associated with fine-grained sediments and TOC. Sieving could also reduce concentrations of volatile constituents and concentrations of hydrophobic chemicals could be reduced through sorption to the sieve itself. Increasing the relative concentrations of fine grained materials might also negatively affect test organisms sensitive to this constituent. It was assumed that sieving increased concentrations of the constituents that were causing toxicity in the untreated sample since organic and metal contaminants are associated with the fine-grained fractions and TOC. Effects of sieving on sample toxicity can be assessed by comparing toxicity of the sample before and after sieving, and by conducting grain size, TOC and chemical analyses on sieved and un-sieved samples.

3.4.6 Factors Associated with Interstitial Water Testing

A number of artifacts may occur as a result of the extraction and testing of interstitial water, and these should be considered as part of interpretation of IW TIE results, particularly when comparing them to solid-phase TIEs. These considerations are discussed in detail in a recent EPA sediment TIE guidance document (U.S. EPA, in press), and are summarized here. One example is that IW is often extracted from anoxic sediments, and once extracted may require aeration to accommodate toxicity test species. This may result in alteration of the redox state of the sample and may affect chemical bioavailability, relative to that in solid-phase sediment. For example, metal bioavailability may be altered through formation of precipitates as a result of exposure to oxygen, and metals and organics may be influenced by changes in IW sample pH.

An additional consideration concerns loss of hydrophobic chemicals in IW tests due to sorption to test containers. Interstitial water tests often are conducted in small volume test containers because of the difficulty of extracting adequate IW volumes. As a consequence, there may be a loss of highly hydrophobic contaminants due to their association with test container walls. For example, Wheelock et al. (2005) found that up to 80% of pyrethroid pesticides may be lost in tests conducted in glass test containers.

In addition to these considerations, exposure of organisms to interstitial water may overestimate the degree of exposure to contaminants that might occur in solid-phase tests. Differences between IW and solid-phase tests may result from test species avoiding contaminants in sediments by escaping to overlying waters, while in IW tests, exposure to chemicals is constant due to lack of refugia. One final consideration is that differences in response between solid-phase and IW tests may reflect differences in contaminant uptake routes. Because the majority of sediment and organic carbon is removed from IW via centrifugation, IW tests may under-represent contaminant uptake via particle ingestion.

3.5 Future Research Needs

3.5.1 TIE Methods for Organic Chemicals

A key step in the Phase II TIE process described in this report involves solvent elution of chemicals from resins such as Ambersorb, or from the HLB solid-phase extraction column used to treat interstitial water. After pumping solvent through the extraction media, the solvents are spiked into clean water to verify that toxic concentrations of chemicals were recovered from the sample. Concentrations of organic chemicals are also measured in the eluates, and these may be compared to their respective toxicity thresholds (e.g., LC50s) to identify specific chemicals responsible for toxicity. Inconsistent elution of toxic concentrations of chemicals was noted in the solid-phase sediment experiments using Ambersorb 563, and in interstitial water experiments using the HLB column. This occurred in experiments using sediments spiked with single chemicals and chemical mixtures, and in the TIEs of the ambient samples.

Because Phase II TIE procedures for Ambersorb were largely undeveloped prior to this project, methanol elution procedures recommended for SPE columns used in effluent TIEs were used (e.g., C18 SPE elution, U.S. EPA 1991, 1993, 1996). Ambersorb elution results using methanol and acetone also were compared, and although better results with acetone were sometimes observed, the results suggested these procedures required further research. Based on conversations with chemists involved in this project, the research team followed the manufacturer's generic procedures using methanol for eluting the HLB column used in the interstitial water TIEs (Waters Corp., 2001). Ankley et al. (1991) discussed problems associated with using methanol to elute less polar organic compounds from C18 columns in interstitial water TIEs, and noted unpredictable recoveries using this solvent. These authors recommended using an increasingly non-polar methylene chloride/methanol elution series to extract more non polar compounds. In their procedure, the SPE column is first eluted three times with 100% methanol, followed by three elutions with 100% methylene chloride. The methylene chloride and methanol extracts are combined, and because methylene chloride is more volatile, it can be sparged using nitrogen. This results in a solvent transfer from methylene chloride to methanol of any remaining chemicals not eluted by the methanol. The research group has not evaluated this procedure with either Ambersorb or the HLB solid-phase extraction column, but assumes this approach would result in better recovery of organic chemicals bound to these media.

The research group recommends these procedures be evaluated using Ambersorb and other resins used to extract organic chemicals in solid-phase TIEs and to the HLB and other SPE columns used in interstitial water TIEs. In these experiments, the Ambersorb was eluted with acetone then subjected to HPLC fractionation with methanol. As discussed in the section on ambient sample TIEs, it is likely the acetone elution was not completely efficient at solubilizing all organic chemicals sorbed to the resin, particularly with samples contaminated by highly hydrophobic compounds such as pyrethroid pesticides. Further loss may have occurred during the HPLC process, because this used methanol, a more polar solvent. Additional loss probably occurred when the HPLC methanol fractions were spiked into water then re-solubilized in iso-octane; this solvent exchange step led to considerable sample dilution prior to analysis with gas chromatography.

As previously discussed, the methods need to be amended to improve elution of chemicals from the solid-phase extraction media, and to alter methods to minimize loss of chemicals through sorption and dilution during the sample preparation steps, prior to analysis. The research group is now in consultation with the analysts to amend the Ambersorb elution and eluate chemistry procedures, and is using revised methods to re-analyze eluates of the SZ Ambersorb samples used in the ambient sample TIEs. As noted in the discussion of the ambient sample TIEs, the research team was able to detect toxic concentrations of pyrethroid pesticides in acetone eluates from several of the ambient samples when the acetone eluate was injected directly into the gas chromatograph, rather than being subjected to the solvent exchange process.

An additional problem observed in the solid-phase TIEs with Ambersorb was incomplete reduction of toxicity in highly toxic samples. In these experiments, the largest Ambersorb volume used was 20%, and this was mixed with sediment for no longer than 24 hours. While this procedure often reduced toxicity in lower dilutions of the ambient sediment samples, it was less effective in higher concentrations of severely toxic sediments (e.g., Consolidated Slip). It is preferable to use undiluted sediment in TIEs, both to avoid overlooking toxicologically important contaminants that might be minimized at lower dilutions, and to reduce other confounding factors introduced by dilution. The research group recommends that additional research be devoted to increasing the effectiveness of carbonaceous resins in solid-phase TIEs. This should include determination of the optimal volume of Ambersorb for use in solid-phase TIEs, and to determine appropriate equilibration times necessary to maximize sorption of chemicals from sediments highly contaminated by non-polar organics. This is particularly important given that non-polar organic compounds may be bound by different carbon sources in sediments, and that desorption rates vary depending on the nature of the carbon present in ambient samples (e.g., soot or black carbon-like materials vs. natural organic matter). A number of papers have suggested that desorption rates vary from hours, days, and months depending on the chemical contaminants and carbon constituents present in bedded sediments (e.g., Cornelissen et al., 2000).

An incomplete removal of non-polar organic compounds using the HLB column in interstitial water TIEs was also observed. With these samples, complete reduction of toxicity by HLB extraction was observed in lower dilutions of interstitial water, but column breakthrough occurred at higher concentrations. A 6 ml HLB column packed with 500 mg of sorbent material was used in these experiments. HLB columns containing greater resin masses are available from the manufacturer, and the increased sorbent surface area should be more effective at binding higher concentrations of non-polar organic chemicals in IW samples. These columns are being investigated for application to interstitial waters contaminated with high concentrations of pyrethroid pesticides, and similar experiments should be applied to other classes of contaminants common to sediments. As in sold-phase sediment TIEs, the goal with interstitial waters is to develop TIE methods for complete extraction of undiluted sample in order to avoid incomplete characterizations that may occur with diluted samples.

3.5.2 Additional TIE Methods for Non-Polar Organics

While recent studies have shown that measuring tissue residues is a useful approach to estimate chemical exposure, its application in sediment TIEs requires more research. Tissue chemicals were measured in amphipods as part of the weight-of-evidence in the TIEs conducted on Switzer Creek (SZ) sediments, after the Phase I TIE of SZ suggested both metals and organics may have played a role in toxicity. In this study, SZ sediments were diluted to a concentration expected to cause 50% mortality in order to provide sufficient living amphipod tissue for chemical analysis. The sediment was then subjected to treatment with SIR-300 to reduce bioavailability of cationic metals, and Ambersorb to bind non-polar organics. Concentrations of metals and organics were then measured in amphipod tissues from these treatments and controls.

Numerous metals were detected in all amphipod tissues, and there was no difference in tissue metal concentrations between the two TIE treatments and the control animals. This suggests that metals may not have been playing a role in toxicity of SZ sediment. Trace concentrations of DDE and oxadiazon were the only organic chemicals detected in amphipod tissues. Relatively high concentrations of chlordane and numerous PAHs were measured in SZ sediment, but no chlordane was detected in the tissues. Unfortunately, PAHs were not analyzed due to inadequate sample size. Concentrations of several pyrethroids exceeded amphipod LC50 values in SZ sediment, but these too were not detected in the tissues.

To determine whether there were sufficient concentrations of organic chemicals to account for toxicity of SZ sediment, the researchers intended to compare tissue concentrations to specific critical tissue LC50s for individual chemicals from the literature. In addition, PAHs and organochlorine pesticide concentrations were intended to be compared to established threshold values for mortality due to cumulative non-polar narcosis. For example, the range of tissue concentrations where non-polar narcosis is expected for fish has been reported to be 0.2 – 0.8 μmol/g (McCarty and Mackay, 1993), although Landrum et al. (2005) found a lower threshold of response for *H. azteca* exposed to DDE. Because so few organics were detected, these comparisons could not be made.

In order to apply tissue residues in TIE analyses, sufficiently low detection limits are necessary, and this requires a relatively large amphipod tissue mass. The researchers recovered live animals in the current experiments, and this yielded approximately 3.3 to 7.0 g of amphipod tissue. Once extracted these were analyzed for organochlorine pesticides, PCBs, and pyrethroid pesticides, and the reporting limits for these chemicals based on the sample sizes ranged from 5 to 10 ng/g fresh wt. for organochlorine pesticides, and 1.0 ng/g for PCBs. The reporting limits for pyrethroids were considerably higher (5 – 20 ng/g). There was insufficient tissue to measure concentrations of PAHs (in addition to the pesticides and PCBs) so emphasis was placed on pesticide analyses, based on the TIE and sediment chemistry evidence. Landrum et al. (2005) recently reported measuring DDE in live and dead amphipods after water-only exposures. Although these authors found differences in critical body residue effect concentrations between live and dead amphipods, their results suggest tissue analyses need not be restricted to live animals only. This could yield greater tissue mass and help lower detection limits in tissue analyses conducted in conjunction with TIEs. It may not be practical to recover dead amphipods after 10-d whole sediment exposures due to their rapid degradation, but measures of tissue concentrations in dead animals may be possible if shorter whole sediment exposure times are used, or by using interstitial water exposures. These approaches should be investigated further to determine the utility of this technique in the TIE process. Also, although the evidence suggests pyrethroids were present in sufficient concentrations in the SZ sediment to account for toxicity, it is not clear that pyrethroid pesticides accumulate to a sufficient degree to allow their detection in amphipod tissues. Further research is needed to determine the degree to which pyrethroids accumulate in amphipod tissues, and to develop lower detection limits for tissue residues of these pesticides to apply these analyses to the TIE process.

3.5.3 Tissue Analysis

This study emphasized current U.S. EPA sediment TIE methods, but additional procedures have recently been reported for determining the bioavailable fraction of organic chemicals in sediments, and these are applicable to sediment TIEs. Heinis et al. (2004) presented a method for incorporating semi-permeable membrane devices (SPMD) in sediment TIEs to assess toxicity by organic toxicants. This approach was developed in response to perceived technical problems involved in isolating carbonaceous resins from solid-phase samples (the approach used in the current project), and in response to practical constraints imposed by the large volumes of sample material required for interstitial water TIEs. In the SPMD-based approach, organic chemicals are extracted from sediment using a combination of solvents. This extract is then mixed into triolein; the mixture is placed into low density polyethylene tubing, and the tubing ends are sealed to provide a SPMD. The triolein and extract-filled SPMD is then placed in water, allowed to equilibrate, and toxicity tests are conducted in this system using appropriate test species. In theory, this approach can be used to model the equilibration of organic chemicals that occurs naturally in sediments, and provides a means of isolating and fractionating organic chemicals responsible for sediment toxicity. After experiments to optimize the system and evaluate possible artifacts, the utility of the SPMD-based approach to TIEs was investigated using sediments spiked with four model compounds and with field-collected sediments. The results showed that this approach can be used to isolate organic toxicants and re-introduce them to a water test system, and the concentrations of the contaminants in this system approximated those in the original sediment (Heinis et al., 2004). This system provides an alternative approach to isolate non-polar organics from solid-phase samples, and may be most useful as an additional line-of-evidence to the solid-phase and interstitial water treatments described in this work. Future research should include comparisons of the efficacy of these methods for application to sediment TIEs, particularly to determine whether conclusions regarding chemicals responsible for toxicity are similar when the different approaches are employed.

3.5.4 TIE Methods for Cationic Metals

Residual toxicity observed in copper-spiked sediments and interstitial waters after treatment with SIR-300 and the Supelco cation column suggest TIEs using these extraction media also require additional refinement. As suggested above for Ambersorb, the research group recommends experiments be conducted to investigate the combination of optimal resin concentrations and equilibrations times in order to maximize metal binding in solid-phase TIEs. Also noted was an incomplete removal of IW toxicity with the cation column, presumably due to column breakthrough. This should be addressed by comparing the binding capacities of columns having greater sorbent volumes. It was also noted in the ambient sample Phase I TIEs that SIR-300 reduced toxicity in samples where the cause of toxicity was clearly an organic. Although Burgess et al. (2000) did not see ameliorative effect on the toxicity of endosulfan, additional experiments need to be conducted to determine the binding capacity of SIR-300 for other organic contaminants.

Of particular concern in these experiments was the inability to elute toxic concentrations of copper from the SIR-300 used to treat marine solid-phase samples spiked with copper. For example, there was a high survival of *E. estuarius* in waters spiked with acid eluates from SIR-300, even when total copper concentrations in spiked waters were three to five times the copper LC50 for this species. It is not clear whether this was due to reduced copper bioavailability, or some other factor. More consistent toxicity was observed in TIEs of interstitial water from copper-spiked sediment when eluates from the cation column used to treat interstitial water were spiked into seawater. Research to investigate the discrepancy between the two approaches should be conducted. For example, although total copper concentrations were well above the LC50, it is not clear how much copper was bioavailable in acid eluates from SIR-300, after they were spiked into seawater. One approach to address this would be to measure cupric ion concentrations in these solutions. Cupric ion concentrations also could be measured in solutions spiked with acid eluates from the cation column to compare cation bioavailability using these two methods.

3.6 Sediment TIE Logistics

Conducting sediment TIEs is an iterative process. The amount of time between the first day of the initial test and the final conclusion is dependent on the study question and the resources of the project. Table 3-8 summarizes the procedures used for the sediment and interstitial water TIEs associated with the Consolidated Slip ambient sample used in this study. The start date, duration, and approximate number of technician hours also are listed to provide perspective on a sediment TIE effort. The process follows the sediment TIE decision matrix outlined in Figure 2-1. The number of technician hours includes test initiation, monitoring, termination, and data interpretation.

TIEs conducted with CS sediment and interstitial water spanned approximately six months. Once the results of the initial tests were interpreted, the TIEs on each matrix began simultaneously. As each TIE procedure was completed, decisions were made on what the next procedure would be. In some cases the decision was based on external analyses, such as HPLC fractionation or chemical analysis.

Table 3-8. Procedures and Technician Hours Necessary to Investigate the Causes of Toxicity in Sediment from Consolidated Slip.

Date	Procedure	Duration (days)	Approximate Technician Hours
4/1/05	Initial sediment test with dilution	11	5
4/1/05	Initial interstitial water test with dilution	10	4
6/7/05	Second initial sediment test with dilution	11	5
6/24/05	Sediment TIE with three treatments and blanks	12	8
6/24/05	First Interstitial Water TIE with eleven treatments and blanks	10	18
7/22/05	Elution of solid-phase TIE amendments	10	5
7/29/05	Second Interstitial Water TIE with nine treatments and blanks	10	16
8/19/05	Third Interstitial Water TIE with thirteen treatments and blanks	10	21
10/10/05	Ambersorb Acetone Eluate HPLC Fractionation Test	10	8

Ambient sample TIEs treatments in this study were conducted simultaneously, for the most part. A few interstitial water treatments were conducted sequentially, but also synoptically with other treatments. Another way of conducting interstitial water TIEs is to manipulate the sample one treatment at a time, or conduct sequential treatments (SAIC 2003). Conducting TIEs in this manner could alter the time frame outlined in Table 3-8. Approximate costs of sequential testing are summarized in SAIC (2003).

APPENDIX A

TOXICITY IDENTIFICATION EVALUATION METHODS

Sediment TIE Methods

This project evaluated the effectiveness of four resin/amendments for use in solid-phase sediment TIEs. These were the carbonaceous resin Ambersorb 563, powdered coconut charcoal (PCC), SIR-300, and zeolite. Ambersorb and PCC reduce bioavailability of organic chemicals, while SIR-300 and zeolite reduce metals and ammonia, respectively. Initial experiments with these amendments evaluated a number of different concentrations. As the research progressed, data for Ambersorb, PCC and SIR-300 indicated the optimal concentration for TIEs was a 10% (by-volume) addition of these amendments. A 20% volume addition of zeolite was used.

Initial Toxicity Test

General Approach

The Initial Toxicity Test is the test that determines whether or not a sample is sufficiently toxic to justify a TIE. Initial tests should be conducted using both solid phase and interstitial water test matrices where practical. If possible, solid phase and interstitial water samples should be tested with a dilution series to assess the magnitude of toxicity of both sediment matrices.

Materials

Materials, organisms, and apparatus necessary to conduct toxicity test.

Procedural Overview

The results of the initial test will determine if a TIE is necessary, and the appropriate sediment matrices to be used in the TIE investigation. In cases where both matrices are toxic, TIEs should be performed on both the solid phase and interstitial water. In either case, the magnitude of toxicity should be sufficient to distinguish between baseline results and those of the TIE treatments. Deciding whether a sample is sufficiently toxic to pursue a TIE requires the best professional judgment of the analysts. As a general rule, a minimum of 50% mortality is preferred in undiluted solid-phase sediment before conducting a TIE. Interstitial water should have complete mortality in undiluted sample, or at least two toxic units (U.S. EPA, 1996). These are not strict requirements, but it is more difficult to resolve differences among TIE treatments in samples with low or moderate toxicity. This decision is driven in part by regulatory requirements and on resources available for the study.

Baseline Toxicity Test

General Approach

The Baseline Toxicity Test is conducted on un-manipulated sample. Baseline test results are compared to the initial test to confirm the magnitude of toxicity, and to results of the various TIE treatments. Baseline results can also indicate whether sample toxicity has declined, which can provide clues to the kind of contaminant causing toxicity. The test is generally conducted on undiluted sample, but if the toxic signal is strong enough to overwhelm the TIE treatments, then diluted sample can be used. If a diluted sample is being tested, it should be at concentrations at or near the lowest observed effect concentration (LOEC).

Materials

Materials, organisms, and apparatus necessary to conduct toxicity test.

Procedural Overview

Baseline test design is based on the toxicity produced in the initial test. If a dilution series was tested, the optimal concentration for treatment should be chosen to include the lowest concentrations with complete mortality and a series of concentrations below this. The number and size of the replicates can be reduced to maximize the use of the sample. For most standard U.S. EPA procedures, five 250-mL replicates containing 50 mL sediment, 200 mL overlying water, and five to ten organisms can be used.

The results of the TIE manipulations are compared to the Baseline results to determine if the treatments have significantly decreased or increased toxicity.

SIR-300 Addition

This procedure is adapted from Burgess et al. (2000).

General Approach

SIR-300 (ResinTech, West Berlin, NJ) is a macroporous weak acid cation exchange resin based on the iminodiacetate acid functional group, which has chelating properties for heavy metal ions even in conditions with high calcium concentrations. After preparation, SIR-300 can be mixed into sediment to reduce cationic metal bioavailability (Burgess et al., 2000).

Materials

Treatment Preparation – SIR-300, top-loading balance (1 kg capacity), 100 µm Nitex® polycarbonate screen tube, Nanopure® water, 500 mL wide-mouth jar, and polypropylene spoon.

Test Termination – Brass sieves ranging from approximately 250 µm to 450 µm.

Resin Elution – 5-mL syringe, 100 µm Nitex mesh, peristaltic pump and silicone tubing, 1N hydrochloric acid (HCl), 1N sodium hydroxide (NaOH), methanol, ring stand and clamp, and micro-pipettor.

Procedural Overview

SIR-300 is prepared by thoroughly rinsing the resin beads with tap water in a polycarbonate screen tube followed by a final rinse using Nanopure water (Barnstead International, Dubuque, IA, U.S.). Ten percent SIR-300 (wet weight) is added to the sediment in a 500-mL mixing jar. Approximately 250g of resin-amended sediment is sufficient for five, 250 mL replicates each containing approximately 50 mL sediment. Treated sediment is homogenized for 24 h on a sediment roller apparatus, and then loaded into toxicity test exposure beakers. The SIR-300 test is conducted without overlying water renewal. Exposures with *H. azteca* are fed 0.5 mL YCT every other day.

A dilution blank is created to account for the possible reduction of sample toxicity that may occur when diluting sediment with SIR-300. At MPSL a dilution control is created by combining test sediment with the 10% clean control sediment. A SIR-300 blank is created to account for any toxic effects created by adding the resin to test sediment by combining 10% SIR-300 with control sediment. All blanks are equilibrated on the sediment roller as described above.

At test termination the sediment is screened through a series of sieves ranging from 250 to 450 μm mesh to separate the resin from the sediment by using differences in density. Begin with a 500 μm screen to remove organisms, and catch the material that passes through in a tray. Proceed to screen the material in sieves with increasingly smaller mesh size. At each stage, dip the sieve in water and swirl the material. The resin will separate from the sediment. Collect the resin at each stage and pool the resin from all replicates. Using this method it is possible to recover approximately 80-90% of the original 25g added to the sediment. To facilitate this process, test sediment should be dry-sieved through 280 μm mesh before being treated and tested. This creates a smaller sediment size distribution that facilitates sieve-separation of the SIR-300. If the sediment was pre-sieved, a 280 μm mesh sieve is used. If the test sediment is pre-sieved for this treatment, the results should then be compared to baseline sediment that has also been pre-sieved.

SIR-300 that has been collected at test termination can be eluted by loading the resin into a 5-mL syringe and passing 1N hydrochloric acid (HCl) through the resin bed using a peristaltic pump. Prepare a polypropylene syringe by placing a small circle of 100μm Nitex screen inside the syringe over the bottom opening. The syringe is then clamped on a ring stand. Load the syringe with approximately 6 - 7g SIR-300 using a squirt bottle filled with Nanopure water to rinse the resin into the syringe. Record the weight of the resin added to the syringe. Rinse resin thoroughly until the water runs clear. Attach a 50-cm piece of silicone tubing to the syringe opening and pass this through the peristaltic pump. Tubing should be pre-cleaned by pumping 25 mL methanol followed by 25 mL Nanopure at 10 mL per minute. Before elution, remove extra water from resin-filled syringe by continuing to run the pump at 10 mL per minute until dry. Adjust pump flow to 1 mL per minute and slowly add 10 mL 1N HCl to the top of the syringe. Collect the acid eluate in a small beaker. An additional resin column can be prepared for chemical analysis. The acid passed through this column should be stored in a small, well-sealed vial. Note the initial volume in case some is lost to evaporation. Add acid to 100 mL clean dilution water adjusted to the appropriate salinity or conductivity. Neutralize the acid-spiked water with 1N sodium hydroxide (NaOH) as appropriate. Elute SIR-300 sieved from test sediment treatments and blanks. Create an acid control by adding 10 mL to 100 mL clean dilution water and neutralizing.

Resin eluates are tested as water-only exposures. After neutralizing, place ten mL of eluate in each of three to five replicate 10-mL scintillation vials. Depending on the amphipod species being tested, one to five animals can be placed in each vial. Conduct the exposure for up to ten days, or until acceptable results have been achieved.

Ambersorb Addition

This procedure is adapted from Kosian et al. (1999) and West et al. (2001).

General Approach

Ambersorb 563® (Rohm and Haas, Spring House, PA, U.S.) is a carbonaceous, non-polar adsorbant produced by pyrolysis of a highly sulfonated styrene-divinylbenzene macroreticular ion-exchange resin. This resin has a moderate surface area, and when mixed with sediment it is effective at reducing the bioavailability of organic contaminants.

This material is no longer being produced, but there are alternative products developed for the same application. Amberlite XAD4 is currently being tested at MPSL and other laboratories. There are also several other products in the Amberlite series. Although the methods described below were developed using Ambersorb 563, it is assumed they are applicable to other carbonaceous resins of similar size.

Materials

Treatment Preparation – Ambersorb, top-loading balance (1 kg capacity), 100 μm Nitex polycarbonate screen tube, Nanopure water, 500-mL wide-mouth jar, and polypropylene spoon.

Test Termination – Brass sieves ranging from approximately 250 μm to 450 μm.

Resin Elution – 5-mL syringe, 100 μm Nitex mesh, peristaltic pump and silicone tubing, acetone, methanol, ring stand and clamp, and micro-pipettor.

Procedural Overview

Ambersorb is prepared by thoroughly rinsing the resin beads with tap water in a polycarbonate screen tube followed by a final rinse with Nanopure water. Ten percent Ambersorb (wet weight) is added to the sediment in a 500-mL mixing jar. Approximately 250g of resin-amended sediment is sufficient for five, 250-mL replicates each containing approximately 50 mL sediment. Treated sediment is homogenized for 24 h on a sediment roller apparatus, and then loaded into toxicity test exposure beakers. The Ambersorb test is conducted without overlying water renewal. Exposures with *H. azteca* are fed 0.5 mL YCT every other day.

A dilution blank is created to account for the possible reduction of sample toxicity that may occur with addition of Ambersorb. At MPSL a dilution control is created by combining test sediment with the 10% clean control sediment. An Ambersorb blank is created to account for any toxic effects resulting from adding the resin to test sediment by combining 10% Ambersorb with control sediment. All blanks are equilibrated on the sediment roller as described previously.

At test termination the sediment is screened through a series of sieves ranging from 250 to 450 μm mesh to separate the resin from the sediment by using differences in density. Begin with a 500 μm screen to remove organisms, and catch the material that passes through in a tray. Proceed to screen the material in sieves with increasingly smaller mesh size. At each stage, dip the sieve in water and swirl the material. The resin will separate from the sediment. Collect the resin at each stage and pool the resin from all replicates. Using this method it is possible to recover approximately 80-90% of the original 25g added to the sediment. To facilitate this process, test sediment should be dry-sieved through 280 μm mesh before being treated and tested. This creates a smaller sediment size distribution that facilitates sieve-separation of the Ambersorb. If the sediment was pre-sieved, a 280 μm mesh sieve is used. If the test sediment is pre-sieved for this treatment, the results should then be compared to baseline sediment that has also been pre-sieved.

Ambersorb that has been collected at test termination can be eluted by loading the resin into a 5-mL syringe and passing acetone through the resin bed using a peristaltic pump. Prepare a glass or polypropylene syringe by placing a small circle of 100 μm Nitex screen inside the syringe over the bottom opening. The syringe is then clamped on a ring stand. Load the syringe with approximately 6 - 7g Ambersorb using Nanopure squirt bottle to rinse the resin into the syringe. Record the weight of the resin added to the syringe. Rinse Ambersorb thoroughly until the water runs clear. Attach a 50-cm piece of silicone tubing to the syringe opening and pass this through a peristaltic pump. Tubing should be pre-cleaned by pumping 25 mL methanol followed by 25 mL Nanopure at 10 mL per minute. Before elution, remove extra water from Ambersorb-filled syringe by continuing to run the pump at 10 mL per minute until dry. Adjust pump flow to 1 mL per minute and slowly add 10 mL acetone to the top of the syringe. Collect the acetone eluate in a 50-mL glass beaker to facilitate volatilization. Concentrate acetone to 1 mL by placing in a 40 °C water bath located in a fume hood. Once the desired volume is reached, add acetone to 100 mL clean dilution water adjusted to the appropriate salinity or conductivity. An additional resin column can be prepared for chemical analysis. The solvent passed through this column should be stored in a small, well-sealed vial. Note the initial volume in case some is lost to evaporation. Elute Ambersorb sieved from test sediment treatments and blanks. Create an acetone control by adding 10 mL acetone to 50-mL beaker, concentrating it to 1 mL as described, and adding it to 100 mL clean dilution water.

Resin eluates are tested as water-only exposures. Place ten mL of eluate in each of three to five replicate 10-mL scintillation vials. Depending on the amphipod species being tested, one to five animals can be placed in each vial. Conduct the exposure for up to ten days, or until acceptable results have been achieved.

The acetone fraction can be further fractionated using HPLC as described in U.S. EPA 1993. Fractions can be tested using small volumes as previously described.

Powdered Coconut Charcoal Addition (PCC)

This procedure is adapted from Ho et al. (2004).

General Approach

Powdered coconut charcoal (PCC) is pyrolized, activated coconut husk that has been ground to <45 μm (90-96%, Calgon Carbon, Pittsburgh, PA, U.S.; Ho et al., 2004). PCC is added to sediment to reduce the bioavailability of organic contaminants. Because it provides greater surface area, PCC is more efficient than Ambersorb at sorbing contaminants (Ho et al., 2004). The research group has found it useful to include both PCC and Ambersorb in solid-phase TIEs, particularly in sediments where toxicity is not reduced with Ambersorb addition. Relative toxicities of sediments treated with PCC and Ambersorb can provide useful TIE information, and these data are useful when combined with results of Ambersorb eluate tests and chemistry to characterize and identify causes of sediment toxicity.

Materials

Treatment Preparation – PCC, top-loading balance (1 kg capacity), Nanopure® water, vacuum pump and flask, vacuum filter apparatus with 0.45 μm glass fiber filters, 2000-mL Erlenmeyer flask, 500-mL wide-mouth jar, and polypropylene spoon.

Safety Note: PCC particles are very fine and present a safety hazard if inhaled. At MPSL, technicians work with dry PCC in a fume hood or with respirators to avoid dust inhalation.

Procedural Overview

PCC is prepared by hydrating it with an excess of Nanopure water in a 2000-mL Erlenmeyer flask. The hydrated charcoal is then vacuum-filtered to form a damp slurry. Pour approximately 50 mL hydrated charcoal into the filter apparatus and initiate the vacuum. Once most of the water is vacuumed through, remove the charcoal pellet from the filter and place in a beaker. Care should be taken to not over-vacuum the PCC. Charcoal that is too dry will not homogenize with the sediment. The filtration process is repeated several times to prepare enough PCC for the treatment. Ten to 15% PCC (wet weight) is added to the sediment in a 500-mL mixing jar. The PCC threshold for *H. azteca* is 15%. Approximately 300 g of PCC-amended sediment is sufficient for five, 250-mL replicates each containing approximately 50 mL sediment. Treated sediment is homogenized for 24 h on a sediment roller apparatus, and then loaded into toxicity test exposure beakers. The PCC test is conducted without overlying water renewal. Exposures with *H. azteca* are fed 0.5 mL YCT every other day.

A dilution blank is created to account for the possible reduction of sample toxicity that may occur with addition of PCC. At MPSL a dilution control is created by combining test sediment with the appropriate percentage clean control sediment. A PCC blank is created to account for any toxic effects resulting from adding PCC to test sediment by combining the appropriate percentage PCC with control sediment. All blanks are equilibrated on the sediment roller as previously described.

Zeolite Addition

This procedure is adapted from Besser et al. (1998), Burgess et al. (2003), and Burgess et al. (2004).

General Approach

Zeolite (SIR-600; ResinTech, West Berlin, NJ) is an inorganic-based ion exchange resin that preferentially removes ammonium, potassium and cesium from water. It is intended for use in decontamination of low-level liquid radioactive waste and for ammonia removal from potable water, aquarium and aquaculture tanks, and municipal wastewaters. After preparation, zeolite can be mixed into sediment to reduce interstitial and overlying water concentrations of ammonia. In some cases, zeolite can also bind metals (Burgess et al., 2004).

Reduction of ammonia is evaluated by measuring overlying water concentrations of total ammonia and converting it to unionized ammonia using pH, temperature, and salinity (Phillips et al., 1997). Interstitial water concentrations of total ammonia can also be measured.

Materials

Zeolite, top-loading balance (1 kg capacity), 100 μm Nitex polycarbonate screen tube, Nanopure® water, 500-mL wide-mouth jar, and polypropylene spoon.

Procedural Overview

Zeolite is prepared by thoroughly rinsing with tap water in a polycarbonate screen tube, followed by a final rinse with Nanopure water. The amendment should be rinsed until rinsate runs clear. Ten to twenty percent zeolite (wet weight) is added to the sediment in a 500 mL mixing jar. Approximately 250g of zeolite-amended sediment is sufficient for five, 250 mL replicates containing approximately 50 mL sediment. Treated sediment is homogenized for 24 h on a sediment roller apparatus, and then loaded into toxicity test exposure beakers. The zeolite test is conducted without overlying water renewal. Exposures with *H. azteca* are fed 0.5 mL YCT every other day.

A dilution blank is created to account for the possible reduction of sample toxicity that may occur when diluting sediment with zeolite. At MPSL a dilution control is created by combining test sediment with the appropriate percentage of clean control sediment. A zeolite blank is created to account for any toxic effects created by zeolite addition to the test sediment by combining the appropriate percentage zeolite to control sediment. Any blanks are equilibrated on the sediment roller as previously described.

Carboxylesterase Enzyme Addition

This procedure is adapted from Wheelock et al. (2004, 2006).

General Approach

The porcine carboxylesterase enzyme (Sigma-Aldrich, St. Louis, MO, U.S.) hydrolyzes ester-containing compounds such as pyrethroid pesticides to their corresponding acid and alcohol, which generally are not toxic (Wheelock et al., 2004, 2006). Decreased toxicity with the addition of carboxylesterase is diagnostic of toxicity due to pyrethroids. In the MPSL procedure, enzyme is added daily to sediment overlying water to hydrolyze pyrethroids. This treatment also may be used in a combined treatment with addition of piperonyl butoxide (PBO) to determine if both organophosphate pesticides and pyrethroid pesticides are contributing to toxicity (see below). A bovine serum albumin (BSA) protein-addition control is conducted with this treatment to account for reduction of bioavailability due to complexation by the enzyme addition.

Materials

Treatment Preparation – Lyophilized porcine carboxylesterase, balance, Nanopure® water.

Procedural Overview

Lyophilized enzyme powder is added to Nanopure water to make a stock solution to be added directly to the sediment overlying water in the exposure chambers. The amount of enzyme added is based on the volume of the overlying water, and the tolerance of the test organism. *E. estuarius* and *H. azteca* tolerate an enzyme strength of 500x. The enzyme is added based on units of activity. One 'x' of enzyme activity equals 0.0025 units of enzyme per mL of sample, therefore at 500x, 1.25 units are added per mL of overlying water. Enzyme strength is unique for each lot purchased (Wheelock et al., 2004). Carboxylesterase is added to the overlying water on the day of test initiation, six hours prior to the addition of amphipods. This allows for interaction between the enzyme and pyrethroids.

Because the carboxylesterase enzyme may reduce bioavailability of other organic chemicals through sorption to the enzyme, a separate treatment is required to separate esterase activity from detoxification through sorption. Bovine serum albumin (BSA, Sigma-Aldrich) is added to a separate set of replicates to account for chemical sorption due to the addition of a protein to the sample. The BSA is prepared in the same manner as the carboxylesterase.

Piperonyl Butoxide (PBO) Addition

This procedure is adapted from U.S. EPA (1991).

General Approach

Piperonyl butoxide (PBO, Sigma-Aldrich, St. Louis, MO, U.S.) is a metabolic inhibitor used to block the metabolic activation of acetylcholinesterase-inhibiting organophosphate pesticides (Ankley et al., 1991). It is also a potent synergist of pesticide toxicity, because it inhibits their metabolism (Ware, 1989; Kakko et al., 2000). PBO is added to overlying water daily to reduce toxicity caused by organophosphate pesticides and increase toxicity caused by pyrethroids, and perhaps DDT (Brandt et al., 2002).

Materials

Treatment Preparation – Piperonyl butoxide, balance, and Nanopure® water.

Procedural Overview

PBO is added to Nanopure water to make a stock solution to be added directly to the sediment overlying water in the exposure chambers. The amount of PBO added is based on the volume of the overlying water, and the tolerance of the test organism. Tests at MPSL have demonstrated that the amphipods *E. estuarius* and *H. azteca* tolerate a PBO concentration as high as 500 µg/L. This concentration of PBO is added to the sediment overlying water on the day of test initiation two hours prior to the addition of amphipods.

Temperature Reduction

General Approach

Certain pesticides are more toxic at lower temperature. These include Type 1 pyrethroid pesticides and DDT. Testing warm water organisms such as *H. azteca* at colder temperatures can be used to indicate the presence of pyrethroids and DDT when samples demonstrate greater toxicity at lower temperatures (Brown, 1987; Coats et al., 1989; Ware, 1989). This treatment is essentially a baseline exposure conducted at 15°C instead of 23°C. A number of recent studies have indicated that tests of ambient samples at different temperatures provide corroborating evidence of toxicity caused by pyrethroid pesticides (Phillips et al., 2004; Anderson et al., 2006; Weston, 2006).

Combination or Sequential Treatments

General Approach

Two or more treatments can be conducted both separately and in combination to resolve toxicity caused by multiple classes of contaminants. Combination treatments can be conducted simultaneously or sequentially, depending on their nature.

Interstitial Water Tie Treatments

The following procedures were adapted from U.S. EPA (1991, 1993, and 1996), Burgess et al., (1997, 2003, 2004), and Wheelock et al., (2004, 2006).

Because a large amount of sediment is needed to extract a sufficient amount of interstitial water, efforts are made to conserve the amount of sample used for interstitial water TIEs. The volume of interstitial water necessary to conduct a TIE will depend on the number of treatments and concentrations. Typically, a complete suite of treatments and concentrations can be conducted using a total interstitial water volume of approximately 800 to 1400 mL. The amount of interstitial water that can be extracted from a sediment sample will vary depending on the type of sediment. A conservative estimate of sediment containing 10% interstitial water is assumed.

In the interstitial water TIEs described in this report, interstitial water toxicity to amphipods was assessed using 20-mL glass scintillation vials, each containing 10 mL sample. *E. estuarius* were exposed in five replicates each containing one amphipod, and *H. azteca* were exposed in three replicates, each containing 5 amphipods.

Initial Toxicity Test

See solid-phase section.

Baseline Toxicity Test

General Approach

The Baseline Toxicity Test is conducted on un-manipulated sample. Baseline test results are compared to the initial test to confirm the magnitude of toxicity, and to results of the various TIE treatments. Baseline results can also indicate whether sample toxicity has declined, which can provide clues to the kind of contaminant causing toxicity. Interstitial water tests are generally conducted with a dilution series. The concentrations of interstitial water are chosen based on the toxicity observed in the initial test. If necessary, the dilution series of the initial test should be adjusted to adequately bracket the LC50. This allows for accurate determination of the magnitude of toxicity in the sample and facilitates toxic unit calculation. If less than optimal sample volume is available, the number and size of the replicates can be reduced to maximize the use of the sample.

Materials

Materials, organisms, and apparatus necessary to conduct toxicity test.

Procedural Overview

The results of the TIE manipulations are compared to the Baseline results to determine if the treatments have significantly decreased or increased toxicity. There are sufficient replicates to conduct statistical analysis.

Centrifugation

General Approach

The interstitial water used in the TIE process was extracted using a bench top centrifuge (refrigerated to 4 °C at 3200 *g* for 30 minutes). If necessary, additional centrifugation might be used to remove residual particles that might contribute to toxicity. The research group recommends that an additional centrifugation step be used to remove particles prior to treatment of samples with solid-phase extraction columns.

Materials

Treatment Preparation – Refrigerated centrifuge, centrifuge tubes or bottles, and large-volume pipettes.

Procedural Overview

Determine the volume of interstitial water needed for the TIE, and determine if there is sufficient sediment to produce the desired volume. The amount of sediment necessary to produce enough interstitial water varies depending on sediment grain size and moisture content. As a general rule, it is assumed most sediment samples contain approximately 10% interstitial water. Spin time and velocity also will vary depending on sediment type, but sediments are generally centrifuged for 30 minutes at 3200 *g*. Once interstitial water has been centrifuged, pipette the supernatant from the tubes using a large volume pipette (10-20 mL). Avoid fine particles at the sediment water interface. If an additional centrifuge procedure is necessary, re-spin the interstitial water in a fresh set of tubes, and prepare a centrifuge treatment blank by centrifuging control water.

Aeration

General Approach

The sample is aerated to characterize toxicity due to volatile compounds or surfactants (U.S. EPA 1996).

Materials

Treatment Preparation – Oil-free air pump and tubing attached to 1-10 mL pipettes or Pasteur pipettes, and graduated cylinders or Erlenmeyer flasks.

Procedural Overview

Samples should be aerated in a fume hood. Pour the necessary volume of interstitial water and control water into the aeration containers. Place pipettes in containers and adjust air flow to several small bubbles per second. Aerate for one hour.

Air Stripping

General Approach

The air stripping treatment uses increased pH and stirring to reduce ammonia (U.S. EPA 1993).

Materials

Treatment Preparation – Magnetic stir plates, Teflon stir bars, shallow crystallizing dishes, 1N sodium hydroxide, and 1N hydrochloric acid.

Procedural Overview

Pour the necessary volume of interstitial water and control water into the crystallizing dishes. Adjust each solution to pH 10. Stir the contents of the containers for 4 hours using magnetic stir bars. Adjust back to ambient pH and prepare concentrations.

Zeolite Column

General Approach

Zeolite removes ammonia from the interstitial water by presenting a negatively charged surface that allows for the adsorption of positively charged ions. Zeolite is also effective at removing metals from solution (Burgess et al., 2004). For this reason, the zeolite treatment should be used in conjunction with treatments that bind divalent cations. If metal toxicity is suspected, an additional method of ammonia removal volatilization after air stripping combined with pH adjustment (U.S. EPA, 1993).

Reduction of ammonia is evaluated by measuring interstitial water concentrations of total ammonia and converting it to unionized ammonia using pH, temperature, and salinity (Phillips et al., 1997).

Materials

Treatment Preparation – Zeolite, balance, 100 µm screen tube, Nanopure® water, 20-mL syringe, 100 µm mesh, peristaltic pump and silicone tubing, separatory funnel, ring stand and clamp.

Procedural Overview

Fifteen grams of zeolite is prepared by thoroughly rinsing with tap water in a polycarbonate screen tube, followed by a final rinse with Nanopure water. The amendment should be rinsed until rinsate runs clear.

Prepare a 20-mL glass syringe by placing a small circle of 100 µm Nitex screen inside the syringe over the bottom opening. Clamp the syringe to a ring stand and rinse zeolite mixture into the syringe using Nanopure squirt bottle. Rinse the zeolite by pouring Nanopure through the syringe until the rinsate is clear. Attach a 50-cm piece of silicone tubing to the syringe opening and pass this through a peristaltic pump. Tubing should be pre-cleaned by pumping 25 mL methanol followed by 25 mL Nanopure at 10 mL per minute. Before elution, remove extra water from zeolite-filled syringe by continuing to run the pump at 10 mL per minute until dry.

Using a separatory funnel as a sample reservoir, begin pumping control water through the syringe at 2.5 mL per minute, discarding the first five mL. When column runs dry, begin pumping sample through the column discarding the first five mL. Prepare test dilutions from each aliquot using post-column control water and sample.

EDTA

General Approach

EDTA (Disodium Ethylenediaminetetraacetic acid) is an organic chelating agent that preferentially binds with divalent metals, such as copper, nickel, lead, zinc, cadmium, mercury, and other transition metals to form non-toxic complexes (U.S. EPA, 1996). It will not complex with anionic forms of metals such as selenids, chromates and hydrochromates.

Materials

Treatment Preparation – Disodium EDTA·$2H_20$, balance, Nanopure® water, volumetric flask, Teflon-coated stir bar, and micropipetter.

Procedural Overview

Prepare EDTA stock solution by weighing out 2.78 g disodium EDTA into a 100-mL volumetric flask. Fill the flask with Nanopure water and mix with a stir bar until the EDTA is completely dissolved. To achieve a final EDTA concentration of 60 mg/L, add 2.4 µL of stock solution for every mL control water and interstitial water. *E. estuarius* and *H. azteca* are tolerant of this concentration, but if other amphipods are being tested, a range of concentrations should be used, or the optimal concentration should be predetermined. Allow the EDTA to interact with the sample for three hours prior to exposing the amphipods. If necessary, adjust the pH of the control water and interstitial water to that of the ambient sample. Prepare test dilutions and aliquot to exposure chambers.

Cation Solid-Phase Extraction Column

General Approach

In this treatment, sediment interstitial water is pumped through a cation-exchange column to extract cationic metals. Reduction of toxicity in the column rinsate provides evidence to characterize metal cations as the cause of toxicity. The column is then eluted with hydrochloric acid which is spiked into control water as a Phase II TIE step. When combined with chemical analysis of metals, significant toxicity in the eluate can be used to identify the cause of toxicity (U.S. EPA, 1996; Burgess et al., 1997).

Materials

Treatment Preparation – Cation exchange column (e.g., Supelco LC-WCX, 500 mg, 3 mL), 1N HCl, 1N NaOH, methanol, Nanopure water, peristaltic pump and silicone tubing, separatory funnel, ring stand and clamp.

Procedural Overview

At MPSL, the system is constructed by connecting approximately 50 cm of silicone tubing through a peristaltic pump. Clean the tubing by pumping 10 mL 1N HCl through it, followed by 25 mL Nanopure water at 10 mL per minute. Place the column in the ring stand clamp and attach the tubing to the outlet of the column. When using the Supelco LC-WCX solid-phase extraction column, the following procedure is recommended; for other columns, check manufacturer recommendations. The column was prepared by adjusting water flow to 2.5 mL per minute, then pumping 2 mL of methanol through column followed by 6 mL of distilled water. It is important to leave a small amount of liquid in the column after each step.

Using a separatory funnel attached to the ring stand as a sample reservoir, begin pumping control water through the column at 2.5 mL per minute, discarding the first five mL. Do not let the column run dry. Re-prepare column with methanol and distilled water before pumping sample. Begin pumping the interstitial water sample through the column, discarding the first five mL. The 100% rinsate concentration is then diluted as appropriate for toxicity testing using post-column control water.

At MPSL the cation column is eluted by cleaning the tubing and passing 6 mL 1N HCl through column at a flow rate of 0.5 mL per minute. Collect this acid rinsate and re-elute the column a second time. This allows better elution of bound cations from the column. The acid is collected in a container and the eluate sample is adjusted to the original sample volume by adding control water. Prepare a treatment blank by adding a similar volume of clean acid to control water. Neutralize the eluate and the blank with 1N sodium hydroxide. Prepare test dilutions using the reconstituted eluate sample and the treatment blank.

Organic Compound Solid-Phase Extraction Column

General Approach

In this treatment, reverse phase liquid chromatography is applied to extract nonionic organic toxicants from an interstitial water sample. Reduction of toxicity in the column rinsate provides evidence to characterize toxicity caused by organic compounds (U.S. EPA, 1991, 1996). The column is then eluted with methanol or other solvent to determine if chemicals bound by the column are toxic when reconstituted with control water. Phase II treatments include the serial elution of the column with different concentrations of solvent, as well as HPLC fractionation of the extract to further separate contaminants. Toxicity of the separated fractions is combined with chemical analyses to identify specific chemicals responsible for toxicity.

Materials

Treatment Preparation – Organic solid-phase extraction column (e.g., Waters Oasis HLB, 500 mg, 6 mL), methanol, Nanopure water, peristaltic pump and silicone tubing, separatory funnel, ring stand and clamp.

Procedural Overview

Prepare the system by connecting approximately 50 cm of silicone tubing through a peristaltic pump. Clean the tubing by pumping 25 mL Nanopure water through it, followed by 25 mL methanol at 10 mL per minute. Place the column in the ring stand clamp and attach the tubing to the outlet of the column. For the Oasis HLB column, the following procedure is recommended; for other columns, check manufacturer recommendations. Set flow to 1 mL per minute. Prepare column by pumping 3 mL methanol through column, followed by 5 mL Nanopure water.

Using a separatory funnel as a sample reservoir, begin pumping control water through the column at 2 mL per minute, discarding the first five mL. After control water has been pumped, pump the interstitial water sample through the column, discarding the first five mL. The 100% interstitial water rinsate concentration is then diluted with post-column control water, as appropriate, for toxicity testing.

Prepare the HLB column for solvent elution by cleaning the tubing and washing the column with 4 mL Nanopure water at 0.5 mL per minute. Elute the column by pumping 4 mL methanol. Collect the methanol in a small beaker or vial for storage. If solvent is to be stored for analyses or reconstitution, seal tightly in a small vial. Note initial volume in the event of loss to volatilization. Evaporate the methanol to approximately 1-2 mL (at 40 °C in a water bath using gentle swirling). Reconstitute the solvent in control water at the original sample volume. The percentage of solvent added to control water should not exceed 1%. Prepare a methanol blank with a similar percentage of solvent, and prepare test concentrations.

Carboxylesterase Enzyme

General Approach

The porcine carboxylesterase enzyme (Sigma-Aldrich, St. Louis, MO, U.S.) hydrolyzes ester-containing compounds such as pyrethroid pesticides to their corresponding acid and alcohol, which generally are not toxic (Wheelock et al., 2004, 2006). Decreased toxicity with the addition of carboxylesterase is diagnostic of toxicity due to pyrethroids. This treatment may also be used in conjunction with addition of piperonyl butoxide (PBO) to determine if both organophosphate pesticides and pyrethroid pesticides are contributing to toxicity (see below). A bovine serum albumin (BSA) protein-addition control is conducted with this treatment to account for reduction of bioavailability due to complexation by the enzyme addition. Unlike the enzyme, BSA has no activity and serves only as a source of protein that is similar to that of the enzyme (Wheelock et al., 2006).

Materials

Treatment Preparation – Lyophilized porcine carboxylesterase, balance, Nanopure® water.

Procedural Overview

Lyophilized enzyme powder is added to Nanopure water to make a stock solution to be added directly to the interstitial water. The amount of enzyme added is based on the volume of the sample and control, and the tolerance of the test organism. *E. estuarius* and *H. azteca* tolerate an enzyme strength of 500x. The enzyme is added based on units of activity. One 'x' of enzyme activity equals 0.0025 units of enzyme per mL of sample, therefore at 500x, 1.25 units are added per mL of sample and control. Enzyme strength is unique for each lot purchased (Wheelock et al., 2004). Carboxylesterase is added to the interstitial water on the day of test initiation, two hours prior to the addition of amphipods. This allows for interaction between the enzyme and pyrethroids.

Because the carboxylesterase enzyme may reduce bioavailability of other organic chemicals through sorption to the enzyme, a separate treatment is required to separate esterase activity from detoxification through sorption. Bovine serum albumin (BSA, Sigma-Aldrich) is added to a separate set of replicates to account for chemical sorption due to the addition of a protein to the sample. The BSA is prepared in the same manner as the carboxylesterase.

Piperonyl Butoxide (PBO)

This procedure is adapted from U.S. EPA (1991).

General Approach

Piperonyl butoxide (PBO, Sigma-Aldrich, St. Louis, MO, U.S.) is a metabolic inhibitor used to block the metabolic activation of acetylcholinesterase-inhibiting organophosphate pesticides (Ankley et al., 1991). It is also a potent synergist of pesticide toxicity, because it inhibits their metabolism (Ware, 1989; Kakko et al., 2000). PBO is added to overlying water daily to reduce toxicity caused by organophosphate pesticides and increase toxicity caused by pyrethroids, and perhaps DDT (Brandt et al., 2002).

Materials

Treatment Preparation – Piperonyl butoxide, balance, and Nanopure® water.

Procedural Overview

PBO is added to Nanopure water to make a stock solution to be added directly to the interstitial water sample. The amount of PBO added is based on the volume of the sample, and the tolerance of the test organism. Tests at MPSL have demonstrated that the amphipods *E. estuarius* and *H. azteca* tolerate a PBO concentration as high as 500 μg/L. This concentration of PBO is added to the sediment overlying water on the day of test initiation two hours prior to the addition of amphipods.

Temperature Reduction

General Approach

Certain pesticides are more toxic at lower temperature. These include Type 1 pyrethroid pesticides and DDT. Testing warm water organisms such as *H. azteca* at colder temperatures can be used to indicate the presence of pyrethroids and DDT when samples demonstrate greater toxicity at lower temperatures (Brown, 1987; Coats et al., 1989; Ware, 1989). This treatment is essentially a baseline exposure conducted at 15 °C instead of 23 °C. As mentioned above, recent studies have indicated that tests of ambient samples at different temperatures provide corroborating evidence of toxicity caused by pyrethroid pesticides (Phillips et al., 2004; Anderson et al., 2006; Weston, 2006).

Combination or Sequential Treatments

General Approach

Two or more treatments can be conducted in sequence or in combination to resolve toxicity caused by multiple classes of contaminants.

Based on experiences at MPSL, some considerations should be taken when planning sequential treatments:

- Pass sample through columns first to maximize removal of contaminant and toxicity.
- Zeolite and air stripping can create precipitate that can clog columns.

APPENDIX B

TOXICITY AND SELECTED CHEMISTRY DATA TABLES

Key to table terminology:

Treatment	Explanation
CU - Baseline	Copper-spiked reference sediment that serves as the baseline sample
CU (10% SIR-300)	Copper-spiked baseline sample with SIR-300 resin added at 10% (wet weight)
Control (10% SIR-300)	Control (reference) sediment with SIR-300 resin added at 10% (wet weight)
CU (10% Control)	Copper-spiked baseline sample with control sediment added at 10% (wet weight)to test for dilution effects
Elution Control	Clean water with addition of acid to mimic elution of SIR-300
Control	Reference sediment
Measured Copper	Copper concentrations measured in the sample and treatments

AMM = ammonia; CU = copper; FLU = fluoranthene, NON = nonylphenol, and TCB = tetrachlorobenzene.

Table B-1. Solid-phase TIE Results from Ammonia-spiked Sediment with *E. estuarius* (mean percent survival and standard deviation (SD)). AMM Indicates Ammonia-spiked Sediment. Ammonia Values are the Mean of the Initial and Final Concentrations Measured in Sediment Overlying Water. Unionized Ammonia Concentrations are Calculated from Overlying Water Ammonia Concentrations, pH, Temperature, and Salinity.

	Test 1		Total $[NH_3]$	Unionized $[NH_3]$
Treatment	Mean	SD	mg/L	mg/L
AMM - Baseline	20	20	137.8	2.03
AMM (20% Zeolite)	56	38	36.1	0.83
Control (20% Zeolite)	90	12	2.3	0.08
AMM (20% Control)	16	26	94.3	1.73
Control	100	0	5.5	0.21

Table B-2. *E. estuarius* TIE Results from Interstitial Water Extracted from Ammonia-spiked Sediment (mean percent survival and standard deviation (SD) from Day 1). Unionized Ammonia Concentrations are Calculated from Overlying Water Ammonia Concentrations, pH, Temperature, and Salinity.

	Percent Sample											Total	Unionized
	0%		10%		25%		50%		100%		Toxic	$[NH_3]$	$[NH_3]$
Treatment	Mean	SD	Mean	SD	Mean	SD	Mean	SD	Mean	SD	Units	mg/L	mg/L
Baseline	100	0	100	0	100	0	100	0	0	0	1.4	690	1.94
Zeolite Rinsate	100	0	100	0	100	0	100	0	60	55	<1	10	0.07
Air Stripping	100	0	100	0	100	0	100	0	80	45	<1	336	4.7
pH 7	100	0	100	0	100	0	100	0	0	0	1.4	684	1.93
pH 8	100	0	100	0	100	0	20	45	0	0	2.5	860	16.9
pH 9	100	0	40	55	0	0	0	0	0	0	12.0	642	107

Table B-3. Solid-phase TIE Results from Copper-spiked Sediment with *E. estuarius* (mean percent survival and standard deviation (SD)). CU Indicates Copper-spiked Sediment.

	Test 2		Test 3		SIR-300 Elution 1		SIR-300 Elution 2		Test 4		SIR-300 Elution	
Treatment	Mean	SD	Mean	SD	Mean	SD	Mean	SD	Mean	SD	Mean	SD
CU - Baseline	8	11	0	0					56	22		
CU (20% SIR-300)	88	27										
CU (10% SIR-300)	88	11	84	17	80	42	100	0	92	11	80	45
Control (20% SIR-300)	83	29										
Control (10% SIR-300)			88	11	100	0	100	0	92	11	80	45
CU (20% Control)	4	9										
CU (10% Control)			80	14					88	18		
Elution Control					90	32	100	0			100	0
Control	100	0	76	22	90	32	100	0	92	11	100	0
Measured Copper			2035 mg/kg						1610 mg/kg		112 mg/L	

Table B-4. Method Evaluation Results for *E. estuarius* TIE Treatments Conducted with Interstitial Water Extracted from Copper-spiked Sediment (mean percent survival and standard deviation (SD) from Day 4). Concentration of Total Copper in Interstitial Water is 51.65 mg/L.

	Percent Sample								
	0%		25%		50%		100%		Toxic
Treatment	Mean	SD	Mean	SD	Mean	SD	Mean	SD	Units
Baseline	100	0	100	0	40	55	0	0	2.1
Filtration	100	0	60	55	40	55	0	0	2.8
EDTA 1	80	45	80	45	80	45	0	0	1.4
EDTA 2	100	0	100	0	100	0	0	0	1.4
STS	100	0	60	55	80	45	0	0	1.6
Cation Rinsate	100	0	100	0	60	55	0	0	1.9
Cation Eluate	100	0	100	0	100	0	100	0	<1
pH 7	80	45	40	55	40	55	0	0	3.0
pH 8	100	0	100	0	60	55	0	0	1.9
pH 9	100	0	80	45	100	0	0	0	1.5

Table B-5. Extraction Efficiency Results for *E. estuarius* TIE Treatments Conducted with Interstitial Water Extracted from Copper-spiked Sediment (mean percent survival and standard deviation (SD) from Day 10).

	Percent Sample									
	0%		25%		50%		100%		Toxic	[Cu]
Treatment	Mean	SD	Mean	SD	Mean	SD	Mean	SD	Units	mg/L
Baseline	100	0	0	0	0	0	0	0	8.0	64.4
Cation Rinsate	80	45	80	45	80	45	0	0	1.4	43.6
Cation Eluate	80	45	100	0	20	45	40	55	2.4	34.7

Table B-6. Solid-phase TIE Results from Fluoranthene-spiked Sediment with *E. estuarius* (mean percent survival and standard deviation (SD)). FLU Indicates Fluoranthene -spiked Sediment.

	Test 3		Ambersorb Elution		Test 4		Ambersorb Elution		Test 5		Ambersorb Elution	
Treatment	Mean	SD	Mean	SD	Mean	SD	Mean	SD	Mean	SD	Mean	SD
FLU - Baseline	0	0			0	0			28	18		
FLU (10% Ambersorb)	88	18	70	48	96	9	0	0	100	0	20	45
Control (10% Ambersorb)	93	10	50	30	74	28	70	48	100	0	100	0
FLU (10% PCC)	88	11			88	18						
Control (10% PCC)	96	9			88	11						
FLU (10% Control)	56	17			4	9			52	30		
Elution Control			70	48			100	0			100	0
Control	85	19	90	32	64	43	100	0	100	0	100	0
Measured Fluoranthene					50.0 mg/kg				20.6 mg/kg		241 μg/L	

Table B-7. Method Evaluation Results for *E. estuarius* TIE Treatments Conducted with Interstitial Water Extracted from Fluoranthene-spiked Sediment (mean percent survival and standard deviation (SD) from Day 4). Fluoranthene was Not Detected in Interstitial Water (detection limit of 1 mg/L).

	Percent Sample								
	0%		25%		50%		100%		Toxic
Treatment	Mean	SD	Mean	SD	Mean	SD	Mean	SD	Units
Baseline	100	0	60	55	0	0	0	0	3.6
Filtration	100	0	100	0	80	45	40	55	1.2
Aeration	100	0	100	0	20	45	0	0	2.5
HLB Rinsate	100	0	80	45	100	0	60	55	<1
HLB Eluate	60	55	0	0	0	0	0	0	8.0
C18 Rinsate	100	0	80	45	80	45	60	55	<1
C18 Eluate	80	45	100	0	60	55	80	45	<1
pH 7	100	0	0	0			0	0	8.0
pH 8	100	0	0	0			0	0	8.0
pH 9	60	55	0	0			0	0	8.0

Table B-8. Extraction Efficiency Results for *E. estuarius* TIE Treatments Conducted with Interstitial Water Extracted from Fluoranthene-spiked Sediment (mean percent survival and standard deviation (SD) from Day 4).

	Percent Sample									
	0%		25%		50%		100%		Toxic	[FLU]
Treatment	Mean	SD	Mean	SD	Mean	SD	Mean	SD	Units	μg/L
Baseline	100	0	100	0	80	45	0	0	1.6	73.2
HLB Rinsate	100	0	100	0	100	0	60	55	<1	3.9
HLB Eluate	80	45	100	0	100	0	100	0	<1	1.4

Table B-9. Solid-phase TIE Results from Nonylphenol-spiked Sediment with *E. estuarius* (mean percent survival and standard deviation (SD)). NON Indicates Nonylphenol-spiked Sediment.

	Test 1		Test 2		Test 3		Ambersorb Elution		Test 4		Ambersorb Elution	
Treatment	Mean	SD	Mean	SD	Mean	SD	Mean	SD	Mean	SD	Mean	SD
NON - Baseline	0	0	0	0	0	0			0	0		
NON (10% Ambersorb)			80	20					20	14	0	0
NON (5% Ambersorb)	32	18	84	9	48	23	0	0				
Control (10% Ambersorb)			88	18					92	18	100	0
Control (5% Ambersorb)	60	37			96	9	100	0				
NON (20% PCC)	96	9	88	18								
NON (10% PCC)			96	9	88	18						
Control (20% PCC)	96	9	80	0								
Control (10% PCC)					88	18						
NON (20% Control)	8	18	4	9								
NON (10% Control)					0	0			0	0		
NON (5% Control)	0	0										
Elution Control							100	0			80	45
Control	96	9	96	9	93	10	100	0	92	11	80	45
Measured Nonylphenol			99 mg/kg		108 mg/kg				215 mg/kg		33 mg/L	

Table B-10. Method Evaluation Results for *E. estuarius* TIE Treatments Conducted with Interstitial Water Extracted from Nonylphenol-spiked Sediment (mean percent survival and standard deviation (SD) from Day 4). Concentration of Nonylphenol in Interstitial Water was Not Measured.

	Percent Sample								
	0%		25%		50%		100%		Toxic
Treatment	Mean	SD	Mean	SD	Mean	SD	Mean	SD	Units
Baseline	100	0	100	0	100	0	0	0	1.4
Filtration	100	0	100	0	100	0	40	55	1.1
Aeration	100	0	100	0	40	55	0	0	2.1
C18 Rinsate	100	0	100	0	100	0	60	55	<1
C18 Eluate	100	0	100	0	100	0	100	0	<1
pH 7	100	0	0	0	0	0	0	0	8.0
pH 8	80	45	0	0	0	0	0	0	8.0
pH 9	100	0	20	45	0	0	0	0	6.4

Table B-11. Extraction Efficiency Results for *E. estuarius* TIE Treatments Conducted with Interstitial Water Extracted from Nonylphenol-spiked Sediment (mean percent survival and standard deviation (SD) from Day 10).

	Percent Sample									
	0%		25%		50%		100%		Toxic	[NON]
Treatment	Mean	SD	Mean	SD	Mean	SD	Mean	SD	Units	µg/L
Baseline	100	0	40	55	0	0	0	0	4.8	350
HLB Rinsate	100	0	80	45	60	55	0	0	1.9	10
HLB Eluate	60	55	40	55	20	45	0	0	2.8	292

Table B-12. Solid-phase TIE Results from Tetrachlorobenzene-spiked Sediment with *E. estuarius* (mean percent survival and standard deviation (SD)). TCB Indicates Tetrachlorobenzene-spiked Sediment.

	Test 1		Test 2		Ambersorb Elution		Test 3		Ambersorb Elution	
Treatment	Mean	SD	Mean	SD	Mean	SD	Mean	SD	Mean	SD
TCB - Baseline	0	0	0	0			4	9		
TCB (10% Ambersorb)			100	0	80	42	88	18	0	0
TCB (5% Ambersorb)	88	11	64	17						
Control (10% Ambersorb)			97	7	100	0	100	0	80	45
Control (5% Ambersorb)	96	9								
TCB (20% PCC)			88	11						
TCB (10% PCC)			48	18						
Control (20% PCC)			96	9						
TCB (20% Control)			8	11						
TCB (10% Control)							12	18		
TCB (5% Control)	4	9								
Elution Control					90	32			100	0
Control	100	0	92	11	90	32	96	9	100	0
Measured Tetrachlorobenzene			402.1 mg/kg				60.5 mg/kg		6.9 mg/L	

Table B-13. Method Evaluation Results for *E. estuarius* TIE Treatments Conducted with Interstitial Water Extracted from Tetrachlorobenzene-spiked Sediment (mean percent survival and Standard Deviation (SD) from Day 4). Concentration of Tetrachlorobenzene in Interstitial Water is 1.42 mg/L.

	Percent Sample						
	0%		50%		100%		Toxic
Treatment	Mean	SD	Mean	SD	Mean	SD	Units
Baseline	100	0	100	0	0	0	1.4
Filtration	100	0	100	0	80	45	<1
Aeration	100	0	100	0	100	0	<1
C18 Rinsate	100	0	100	0	60	55	<1
C18 Eluate	80	45	100	0	100	0	<1
pH 7	100	0	0	0	0	0	4.0
pH 8	100	0	0	0	0	0	4.0
pH 9	80	45	0	0	0	0	4.0

Table B-14. Extraction Efficiency Results for *E. estuarius* TIE Treatments Conducted with Interstitial Water Extracted from Tetrachlorobenzene -spiked Sediment (mean percent survival and standard deviation (SD) from Day 10).

	Percent Sample									
	0%		25%		50%		100%		Toxic	[TCB]
Treatment	Mean	SD	Mean	SD	Mean	SD	Mean	SD	Units	μg/L
Baseline	100	0	100	0	0	0	0	0	2.8	1722
HLB Rinsate	100	0	100	0	80	45	0	0	1.6	6.9
HLB Eluate	80	45	80	45	100	0	60	55	<1	137.4

Table B-15. Solid-phase TIE Results from Ammonia-spiked Sediment with *H. azteca* (mean percent survival and standard deviation (SD)). AMM Indicates Ammonia-spiked Sediment. Ammonia Values are the Mean of the Initial and Final Concentrations Measured in Sediment Overlying Water. Unionized Ammonia Concentrations are Calculated from Overlying Water Ammonia Concentrations, pH, Temperature, and Salinity.

	Test 1		Total [NH_3]	Unionized [NH_3]	Test 2		Total [NH_3]	Unionized [NH_3]
Treatment	Mean	SD	mg/L	mg/L	Mean	SD	mg/L	mg/L
AMM - Baseline	8	8	307	6.85	0	0	90.2	6.51
AMM (20% Zeolite)	92	4	182	3.12	88	11	54.3	4.68
Con (20% Zeolite)	98	4	1.1	0.18	96	9	3.4	0.51
AMM (20% Control)	16	9	277	7.16	12	18	82.6	6.22
Control	100	0	1.7	0.24	88	18	8.2	1.03

Table B-16. *H. azteca* TIE Results from Interstitial Water Extracted from Ammonia-spiked Sediment (mean percent survival and standard deviation (SD) from Day 4). Unionized Ammonia Concentrations are Calculated from Overlying Water Ammonia Concentrations, pH, Temperature, and Salinity.

	Percent Sample											Total	Unionized
	0%		10%		25%		50%		100%		Toxic	[NH_3]	[NH_3]
Treatment	Mean	SD	Mean	SD	Mean	SD	Mean	SD	Mean	SD	Units	mg/L	mg/L
Baseline	100	0	100	0	13	12	0	0	0	0	5.7	1096	12.8
Zeolite Rinsate	93	12	93	12	100	0	80	0	0	0	1.6	567	14.3
Air Stripping	100	0	100	0	100	0	100	0	93	12	<1	290	7.32
pH 7	100	0	67	58	0	0	0	0	0	0	7.9	1032	5.43
pH 8	100	0	20	35	0	0	0	0	0	0	16.0	1045	42.1
pH 9	53	42	0	0	0	0	0	0	0	0	20.0	1024	303

Table B-17. Solid-phase TIE Results from Copper-spiked Sediment with *H. azteca* (mean percent survival and standard deviation (SD)). CU Indicates Copper-spiked Sediment.

	Test 1		Test 3		SIR-300 Elution		Test 4		SIR-300 Elution	
Treatment	Mean	SD	Mean	SD	Mean	SD	Mean	SD	Mean	SD
CU - Baseline	0	0	0	0			2	4		
CU (20% SIR-300)	56	32								
CU (10% SIR-300)			62	23	0	0	76	11	13	12
Control (20% SIR-300)	6	13								
Control (10% SIR-300)			82	15	100	0	100	0	82	17
CU (20% Control)	8	4	0	0						
CU (10% Control)							6	9		
Elution Control					87	12			13	12
Control	100	0	76	13	100	0	96	5	100	0
Measured Copper			1065 mg/kg				753 mg/kg		50.2 mg/L	

Table B-18. Method Evaluation Results for *H. azteca* TIE Treatments Conducted with Interstitial Water Extracted from Copper-spiked Sediment (mean percent survival and standard deviation (SD) from Day 4). Concentration of Copper in Interstitial Water is 12.65 mg/L.

	Percent Sample								
	0%		25%		50%		100%		Toxic
Treatment	Mean	SD	Mean	SD	Mean	SD	Mean	SD	Units
Baseline	93	12	49	15	38	4	13	12	3.3
Filtration	87	12	33	12	53	23	33	31	4.0
EDTA 1	100	0	100	0	93	12	93	12	<1
EDTA 2	87	12	100	0	100	0	88	11	<1
STS 1	47	12	33	31	27	12	47	23	NA
STS 2	33	23	12	11	40	20	34	28	NA
Cation Rinsate	100	0	100	0	93	12	93	12	<1
Cation Eluate	87	12	27	12	20	20	0	0	5.6
pH 7	93	12	0	0	0	0	0	0	8.0
pH 8	100	0	7	12	0	0	0	0	7.5
pH 9	100	0	0	0	0	0	0	0	8.0

Table B-19. Extraction Efficiency Results for *H. azteca* TIE Treatments Conducted With Interstitial Water Extracted from Copper-spiked Sediment (mean percent survival and standard deviation (SD) from Day 7).

	Percent Sample									
	0%		25%		50%		100%		Toxic	[Cu]
Treatment	Mean	SD	Mean	SD	Mean	SD	Mean	SD	Units	mg/L
Baseline	80	20	7	12	7	12	0	0	7.4	14.2
Cation Rinsate	26	31	93	12	100	0	73	12	<1	1.1
Cation Eluate	60	20	0	0	0	0	0	0	8.0	7.8

Table B-20. Solid-phase TIE Results from Fluoranthene-spiked Sediment with *H. azteca* (mean percent survival and standard deviation (SD)). FLU Indicates Fluoranthene-spiked Sediment. Test 3 was Spiked with 120 mg/kg Fluoranthene While the Other Tests were Spiked with 200 mg/kg.

	Test 3		Ambersorb Elution		Test 4		Ambersorb Elution		Test 5		Ambersorb Elution	
Treatment	Mean	SD	Mean	SD	Mean	SD	Mean	SD	Mean	SD	Mean	SD
FLU - Baseline	38	19			0	0			8	18		
FLU (15% Ambersorb)	100	0	73	42	100	0	0	0				
FLU (10% Ambersorb)	92	18	67	12	90	7	0	0	92	8	0	0
Control (15% Ambersorb)	100	0	73	23	100	0	76	8	100	0	87	12
FLU (15% PCC)	64	19			30	14						
FLU (10% PCC)	34	19			34	11						
Control (15% PCC)	38	19			92	11						
FLU (15% Control)	66	22			0	0						
FLU (10% Control)									12	8		
Elution Control			67	12			100	0			100	0
Control	100	0	87	23	100	0	87	12	90	12	100	0
Measure Fluoranthene					220.0 mg/kg				48.7 mg/kg		530 µg/L	

Table B-21. Method Evaluation Results for *H. azteca* TIE Treatments Conducted with Interstitial Water Extracted from Fluoranthene-spiked Sediment (mean percent survival and standard deviation (SD) from Day 10). Concentration of Fluoranthene in Interstitial Water is 1.04 mg/L.

	Percent Sample								
	0%		25%		50%		100%		Toxic
Treatment	Mean	SD	Mean	SD	Mean	SD	Mean	SD	Units
Baseline	87	12	53	23	20	0	7	12	3.2
Filtration	87	12	73	12	33	31	20	35	2.2
Aeration	93	12	100	0	80	35	47	31	1.2
HLB Rinsate	80	20	100	0	40	20	33	23	2.0
HLB Eluate	93	12	93	12	87	23	27	31	1.3
C18 Rinsate	100	0	87	12	60	35	7	12	1.9
C18 Eluate	93	12	87	12	87	23	100	0	<1
pH 7	0	0	0	0	0	0	0	0	NA
pH 8	93	12	0	0	0	0	0	0	8.0
pH 9	93	12	0	0	0	0	0	0	8.0

Table B-22. Extraction Efficiency Results for *H. azteca* TIE Treatments Conducted with Interstitial Water Extracted from Fluoranthene-spiked Sediment (mean percent survival and standard deviation (SD) from Day 10).

	Percent Sample									
	0%		25%		50%		100%		Toxic	[FLU]
Treatment	Mean	SD	Mean	SD	Mean	SD	Mean	SD	Units	µg/L
Baseline	87	12	100	0	100	0	50	10	<1	144.3
HLB Rinsate	40	40	100	0	93	12	93	12	<1	1.9
HLB Eluate	100	0	93	12	87	12	100	0	<1	2.1

Table B-23. Solid-phase TIE Results from Nonylphenol-spiked Sediment with *H. azteca* (mean percent survival and standard deviation (SD)). NON Indicates Nonylphenol-spiked Sediment.

Treatment	Test 1		Test 2		Test 3		Test 4		Ambersorb Elution		Test 5		Ambersorb Elution	
Treatment	Mean	SD	Mean	SD	Mean	SD	Mean	SD	Mean	SD	Mean	SD	Mean	SD
NON - Baseline	0	0	0	0	0	0	0	0			0	0		
NON (15% Ambersorb)							78	11	0	0				
NON (10% Ambersorb)			30	28	0	0					50	29	0	0
NON (5% Ambersorb)	4	9	0	0	0	0								
Control (15% Ambersorb)			98	4	80	7	97	6	87	23				
Control (10% Ambersorb)			98	4	80	7					92	8	93	12
NON (15% PCC)	100	0	50	14	60	34								
NON (10% PCC)	94	5	66	15	68	16								
Control (15% PCC)			86	19	70	18								
NON (15% Control)	68	15	0	0	0	0	0	0						
NON (10% Control)	0	0									0	0		
Elution Control	0	0							100	0			87	12
Control			82	40	96	5	82	4	87	12	92	8	80	35
Measured Nonylphenol					239 mg/kg		314 mg/kg				535 mg/kg		93 mg/L	

Table B-24. Method Evaluation Results for *H. azteca* TIE Treatments Conducted with Interstitial Water Extracted from Nonylphenol-spiked Sediment (mean percent survival and standard deviation (SD) from Day 10). Concentration of Nonylphenol in Interstitial Water was Not Measured.

	Percent Sample										
	0%		10%		25%		50%		100%		Toxic
Treatment	Mean	SD	Mean	SD	Mean	SD	Mean	SD	Mean	SD	Units
Baseline	93	12	40	40	0	0	0	0	0	0	11.4
Filtration	87	12	100	0	60	40	0	0	0	0	3.8
Aeration	87	12	47	23	0	0	0	0	0	0	9.3
C18 Rinsate	87	23	93	12	40	20	0	0	0	0	4.4
C18 Eluate	93	12	93	12	93	12	100	0	73	23	<1
pH 7	53	23	0	0	0	0	0	0	0	0	20.0
pH 8	100	0	0	0	0	0	0	0	0	0	20.0
pH 9	73	12	0	0	0	0	0	0	0	0	20.0

Table B-25. Extraction Efficiency Results for *H. azteca* TIE treatments Conducted with Interstitial Water extracted from nonylphenol-spiked Sediment (mean percent survival and standard deviation (SD) from Day 10).

	Percent Sample									
	0%		25%		50%		100%		Toxic	[NON]
Treatment	Mean	SD	Mean	SD	Mean	SD	Mean	SD	Units	μg/L
Baseline	73	23	0	0	0	0	0	0	8.0	1030
HLB Rinsate	100	0	93	12	73	12	58	23	<1	49.7
HLB Eluate	87	12	0	0	0	0	0	0	8.0	1040

Table B-26. Solid-phase TIE Results from Tetrachlorobenzene-spiked sediment with *H. azteca* (mean percent survival and standard deviation (SD)). TCB Indicates tetrachlorobenzene-spiked Sediment.

	Test 1		Ambersorb Elution		Test 2		Ambersorb Elution		Test 3		Ambersorb Elution	
Treatment	Mean	SD	Mean	SD	Mean	SD	Mean	SD	Mean	SD	Mean	SD
TCB - Baseline	0	0			10	14			10	18		
TCB (10% Ambersorb)					66	43	0	0	98	4	0	0
TCB (5% Ambersorb)	64	13	0	0	72	30						
Con (10% Ambersorb)					96	5	100	0	100	0	67	31
Con (5% Ambersorb)	76	22	80	20								
TCB (20% PCC)	2	4										
TCB (15% PCC)					40	7						
Con (20% PCC)	0	0										
Con (15% PCC)					82	4						
TCB (20% Control)	0	0										
TCB (15% Control)					16	15						
TCB (10% Control)									18	29		
TCB (5% Control)	0	0										
Elution Control			100	0			100	0			100	0
Control	100	0	100	0	100	0	100	0	94	9	87	21
Measured Tetrachlorobenzene					3563 mg/kg				473 mg/kg		99.2 mg/L	

Table B-27. *H. azteca* TIE Results from Interstitial Water Extracted from Tetrachlorobenzene-spiked Sediment (mean percent survival and standard deviation (SD) from Day 10). Concentration of Tetrachlorobenzene in Interstitial Water was 1130 μg/L.

	Percent Sample								
	0%		25%		50%		100%		Toxic
Treatment	Mean	SD	Mean	SD	Mean	SD	Mean	SD	Units
Baseline	93	12	1.00	0	67	31	7	12	1.7
Filtration	80	0	67	23	53	31	20	35	1.5
Aeration	87	12	87	12	47	12	0	0	1.9
HLB Rinsate	100	0	80	20	67	12	7	12	1.8
HLB Eluate	53	46	87	12	93	12	33	31	1.1
pH 7	93	12	0	0	0	0	0	0	8.0
pH 8	100	0	0	0	0	0	0	0	8.0
pH 9	93	12	0	0	0	0	0	0	8.0

Table B-28. Extraction Efficiency Results for *H. azteca* TIE Treatments Conducted with Interstitial Water Extracted from Tetrachlorobenzene-spiked Sediment (mean percent survival and standard deviation (SD) from Day 10).

	Percent Sample									
	0%		25%		50%		100%		Toxic	[TCB]
Treatment	Mean	SD	Mean	SD	Mean	SD	Mean	SD	Units	μg/L
Baseline	93	12	100	0	87	23	80	35	<1	2042
HLB Rinsate	100	0	0	0	100	0	87	23	<1	6.1
HLB Eluate	93	12	100	0	100	0	93	12	<1	283

Table B-29. Solid-phase *E. estuarius* TIE Results from Sediment Spiked with Copper and Fluoranthene (mean percent survival and standard deviation (SD)). CU-FLU Indicates Sediment Spiked with Copper and Fluoranthene. The Combination Treatment Consists of SIR-300, Ambersorb, and Zeolite.

	Test 1	
Treatment	Mean	SD
CU-FLU - Baseline	4	9
CU-FLU (10% SIR-300)	52	11
Control (10% SIR-300)	92	11
CU-FLU (10% Ambersorb)	36	22
Control (10% Ambersorb)	100	0
CU-FLU (20% Zeolite)	4	9
Control (20% Zeolite)	92	11
CU-FLU (40% Combination)	92	11
Control (40% Combination)	92	11
CU-FLU (40% Control)	32	18
CU-FLU (20% Control)	7	10
CU-FLU (10% Control)	8	11
Control	100	0
Measured Copper	741 mg/kg	
Measured Fluoranthene	31.3 mg/kg	

Table B-30. Solid-phase *E. estuarius* TIE and Chemistry Results from Sediment Spiked with Copper and Fluoranthene (mean percent survival and standard deviation (SD)). CU-FLU Indicates Sediment Spiked with Copper and Fluoranthene.

	Test 1		SIR-300 Elution		Ambersorb Elution	
Treatment	Mean	SD	Mean	SD	Mean	SD
CU-FLU - Baseline	12	11				
CU-FLU (10% SIR-300)	92	11	0	0		
CU-FLU (10% Ambersorb)	68	23			40	55
CU-FLU (10% Control)	44	9				
Elution Control			100	0	80	45
Control	100	0	100	0	100	0
Measured Copper	1550 mg/kg		174 mg/L			
Measured Fluoranthene	28.8 mg/kg				180 μg/L	

Table B-31. *E. estuarius* TIE Results from Interstitial Water Extracted from Sediment Spiked with Copper and Fluoranthene (mean percent survival and standard deviation (SD) from Day 10). Interstitial Water Concentration of Copper is 29.5 mg/L and the Concentration of Fluoranthene is 22.3 μg/L.

	Percent Sample										
	0%		10%		25%		50%		100%		Toxic
Treatment	Mean	SD	Mean	SD	Mean	SD	Mean	SD	Mean	SD	Units
Baseline	100	0	60	55	20	45	0	0	0	0	7.9
Filtration	60	55	80	45	80	45	0	0	0	0	2.8
EDTA	100	0	80	45	80	45	40	55	0	0	2.4
Cation Rinsate	80	45	100	0	100	0	60	55	0	0	1.8
Cation Eluate	80	45	80	45	60	55	80	45	60	55	<1
HLB Rinsate	100	0	100	0	60	55	0	0	0	0	3.9
HLB Eluate	100	0	100	0	80	45	80	45	0	0	1.9
Sequential Cation HLB	80	45	100	0	80	45	60	55	0	0	1.9
Sequential Cation Eluate	100	0	100	0	60	55	60	55	40	55	1.4
Sequential HLB Eluate	100	0	100	0	100	0	100	0	20	45	1.3

Table B-32. Extraction Efficiency Results for *E. estuarius* TIE Treatments Conducted with Interstitial Water Extracted from Copper and Fluoranthene-spiked Sediment (mean percent survival and standard deviation (SD) from Day 10).

	Percent Sample										
	0%		25%		50%		100%		Toxic	[Cu]	[FLU]
Treatment	Mean	SD	Mean	SD	Mean	SD	Mean	SD	Units	mg/L	μg/L
Baseline	100	0	60	55	0	0	0	0	3.6	97.4	142.9
Cation Rinsate	100	0	60	55	60	55	0	0	1.8	47.8	ND
Cation Eluate	80	45	20	45	100	0	80	45	<1	28.1	ND
HLB Rinsate	100	0	60	55	20	45	0	0	3.4	76.6	ND
HLB Eluate	100	0	100	0	100	0	100	0	<1	0.9	ND

Table B-33. Solid-phase *E. estuarius* TIE Results from Sediment Spiked with Copper and Ammonia (mean percent survival and standard deviation (SD)). CU- AMM Indicates Sediment Spiked with Copper and Ammonia. Mean Total Ammonia Concentrations are from Sediment Overlying Water. The Combination Treatment Consists of SIR-300, Ambersorb, and Zeolite. Unionized Ammonia Concentrations are Calculated from Overlying Water Ammonia Concentrations, pH, Temperature, and Salinity.

Treatment	Test 1 Mean	Test 1 SD	Total $[NH_3]$ mg/L	Unionized $[NH_3]$ mg/L
CU-AMM - Baseline	0	0	55.4	0.54
CU-AMM (10% SIR-300)	88	11	61.4	0.74
Control (10% SIR-300)	96	9	11.8	0.23
CU-AMM (10% Ambersorb)	4	9	58.8	0.57
Control (10% Ambersorb)	100	0	9.2	0.11
CU-AMM (20% Zeolite)	52	18	34.3	0.33
Control (20% Zeolite)	96	9	6.7	0.09
CU-AMM (40% Combination)	80	14	26.4	0.30
Control (40% Combination)	96	9	3.6	0.06
CU-AMM (40% Control)	68	23	36.3	0.39
CU-AMM (20% Control)	42	15	37.3	0.68
CU-AMM (10% Control)	16	9	57.8	0.55
Control	88	11	11.7	0.16
Measured Copper	1985 mg/kg			

Table B-34. Solid-phase *E. estuarius* TIE and Chemistry Results from Sediment Spiked with Copper and Ammonia (mean percent survival and standard deviation (SD)). CU-AMM Indicates Sediment Spiked with Copper and Ammonia. Mean Total Ammonia Concentrations are from Sediment Overlying Water and SIR-300 Elution. Unionized Ammonia Concentrations are Calculated from Overlying Water Ammonia Concentrations, pH, Temperature, and Salinity.

Treatment	Test 2 Mean	Test 2 SD	Total $[NH^3]$ mg/L	Unionized $[NH^3]$ mg/L	SIR-300 Elution Mean	SIR-300 Elution SD
CU-AMM	0	0	52.9	0.84		
CU-AMM (10% SIR-300)	76	17	56.4	1.14	80	45
CU-AMM (20% Zeolite)	16	17	29.3	0.52		
CU-AMM (20% Control)	12	18	43.1	0.76		
Elution Control					100	0
Control	100	0	1.90	0.06	100	0
Measured Copper	1470 mg/kg				42.1 mg/L	

Table B-35. *E. estuarius* TIE Results from Interstitial Water Extracted from Sediment Spiked with Copper and Ammonia (mean percent survival and standard deviation (SD) from Day 4). Concentration of Copper in Interstitial Water is 33.8 mg/L. Unionized Ammonia Concentrations are Calculated from Overlying Water Ammonia Concentrations, pH, Temperature, and Salinity.

	Percent Sample											Total	Unionized
	0%		10%		25%		50%		100%		Toxic	[NH_3]	[NH_3]
Treatment	Mean	SD	Mean	SD	Mean	SD	Mean	SD	Mean	SD	Units	mg/L	mg/L
Baseline	100	0	40	55	0	0	0	0	0	0	12.0	492	0.30
Filtration	100	0	20	45	0	0	0	0	0	0	15.9	468	0.26
Zeolite	20	45	0	0	0	0	0	0	0	0	NA	130	0.08
Cation Rinsate	80	45	100	0	0	0	0	0	0	0	6.3	488	0.55
Sequential Zeolite EDTA	80	45	100	0	100	0	20	45	0	0	2.4	142	0.08
Seq. Zeolite Cation Rinsate	100	0	100	0	40	55	40	55	0	0	3.5	146	0.22
Seq. Zeolite Cation Eluate	80	45	80	45	100	0	40	55	20	45	1.9	NA	NA
Seq. Air Strip EDTA	80	45	80	45	100	0	00	0	0	0	2.8	276	0.19
Seq. Air Strip Cation Rinsate	80	45	100	0	60	55	20	45	0	0	3.2	294	0.76
Seq. Air Strip Cation Eluate	100	0	100	0	100	0	100	0	60	55	<1	NA	NA

Table B-36. Extraction Efficiency Results for *E. estuarius* TIE Treatments Conducted with Interstitial Water Extracted from Copper and Ammonia-spiked Sediment (mean percent survival and standard deviation (SD) from Day 10). Unionized Ammonia Concentrations are Calculated from Overlying Water Ammonia Concentrations, pH, Temperature, and Salinity.

	Percent Sample										Total	Unionized
	0%		25%		50%		100%		Toxic	[Cu]	[NH_3]	[NH_3]
Treatment	Mean	SD	Mean	SD	Mean	SD	Mean	SD	Units	mg/L	mg/L	mg/L
Baseline	100	0	0	0	0	0	0	0	8.0	35.4	456	0.55
Cation Rinsate	100	0	60	55	0	0	0	0	3.6	9.1	424	2.55
Cation Eluate	100	0	100	0	80	45	100	0	<1	14.7	ND	ND
Zeolite Rinsate	100	0	100	0	0	0	0	0	2.8	22.4	126	0.79

Table B-37. Solid-phase *E. estuarius* TIE Results from Sediment Spiked with Nonylphenol and Ammonia (mean percent survival and standard deviation (SD)). NON-AMM Indicates Sediment Spiked with Nonylphenol and Ammonia. Mean Total Ammonia Concentrations are from Sediment Overlying Water. The Combination Treatment Consists of SIR-300, Ambersorb, and Zeolite. Unionized Ammonia Concentrations are Calculated from Overlying Water Ammonia Concentrations, pH, Temperature, and Salinity.

	Test 1		Total [NH_3]	Unionized [NH_3]
Treatment	Mean	SD	mg/L	mg/L
NON-AMM - Baseline	0	0	80.1	1.48
NON-AMM (10% SIR-300)	0	0	73.3	1.35
Control (10% SIR-300)	88	11	12.5	0.26
NON-AMM (10% Ambersorb)	0	0	72.0	0.96
Control (10% Ambersorb)	68	11	12.5	0.24
NON-AMM (20% Zeolite)	0	0	43.4	0.64
Control (20% Zeolite)	76	9	7.50	0.13
NON-AMM (40% Combination)	64	22	24.8	0.55
Control (40% Combination)	76	26	5.60	0.14
NON-AMM (40% Control)	0	0	28.5	0.52
NON-AMM (20% Control)	0	0	79.7	1.30
NON-AMM (10% Control)	0	0	70.0	1.04
Control	76	17	38.8	0.52
Measured Nonylphenol	162 mg/kg			

Table B-38. Solid-phase *E. estuarius* TIE and Chemistry Results from Sediment Spiked with Copper and Ammonia (mean percent survival and standard deviation (SD)). NON-AMM Indicates Sediment Spiked with Nonylphenol and Ammonia. Mean Total Ammonia Concentrations are from Sediment Overlying Water and Ambersorb Elution. Unionized Ammonia Concentrations are Calculated from Overlying Water Ammonia Concentrations, pH, Temperature, and Salinity.

	Test 2		Total [NH^3]	Unionized [NH^3]	Ambersorb Elution	
Treatment	Mean	SD	mg/L	mg/L	Mean	SD
NON-AMM - Baseline	0	0	51.0	0.83		
NON-AMM (10% Ambersorb)	16	26	45.2	0.81	0	0
NON-AMM (20% Zeolite)	0	0	29.9	0.54		
NON-AMM (20% Control)	0	0	47.3	0.89		
Elution Control					100	0
Control	96	9	1.30	0.04	100	0
Measured Nonylphenol	405 mg/kg				38.5 mg/L	

Table B-39. *E. estuarius* TIE Results from Interstitial Water Extracted from Sediment Spiked with Nonylphenol and Ammonia (mean percent survival and standard deviation (SD) from Day 3). Concentration of Nonylphenol in Interstitial Water is 1470 μg/L. Unionized Ammonia Concentrations are Calculated from Overlying Water Ammonia Concentrations, pH, Temperature, and Salinity.

	Percent Sample											Total	Unionized
	0%		10%		25%		50%		100%		Toxic	$[NH_3]$	$[NH_3]$
Treatment	Mean	SD	Mean	SD	Mean	SD	Mean	SD	Mean	SD	Units	mg/L	mg/L
Baseline	100	0	60	55	0	0	0	0	0	0	8.6	490	1.82
Filter	100	0	100	0	20	45	0	0	0	0	5.4	410	2.15
Zeolite	100	0	100	0	100	0	100	0	0	0	1.4	120	0.95
HLB Rinsate	80	45	100	0	100	0	40	55	0	0	2.1	510	2.93
Sequential Zeolite HLB Rinsate	100	0	100	0	100	0	80	45	0	0	1.6	117	0.84
Seq. Zeolite HLB Eluate	80	45	80	45	100	0	60	55	0	0	1.8	NA	NA
Seq. Air Strip HLB Rinsate	100	0	100	0	100	0	100	0	40	55	1.1	300	3.65
Seq. Air Strip HLB Eluate	80	45	80	45	100	0	40	55	0	0	2.1	NA	NA

Table B-40. Extraction Efficiency Results for *E. estuarius* TIE Treatments Conducted with Interstitial Water Extracted from Nonylphenol and Ammonia-spiked Sediment (mean percent survival and standard deviation (SD) from Day 4). Unionized Ammonia Concentrations are Calculated from Overlying Water Ammonia Concentrations, pH, Temperature, and Salinity.

	Percent Sample										Total	Unionized
	0%		25%		50%		100%		Toxic	[NON]	$[NH_3]$	$[NH_3]$
Treatment	Mean	SD	Mean	SD	Mean	SD	Mean	SD	Units	µg/L	mg/L	mg/L
Baseline	100	0	0	0	0	0	0	0	8.0	2360	356	2.09
HLB Rinsate	100	0	60	55	40	55	0	0	2.8	ND	376	2.63
HLB Eluate	100	0	20	45	0	0	0	0	6.4	1300	ND	ND
Zeolite Rinsate	60	55	100	0	0	0	0	0	2.8	685	110	0.66

Table B-41. Solid-phase *H. azteca* TIE Results from Sediment Spiked with Copper and Fluoranthene (mean percent survival and standard deviation (SD)). CU-FLU Indicates Sediment Spiked with Copper and Fluoranthene. The Combination Treatment Consists of SIR-300, Ambersorb, and Zeolite.

	Test 1	
Treatment	Mean	SD
CU-FLU - Baseline	0	0
CU-FLU (10% SIR-300)	76	17
Control (10% SIR-300)	92	11
CU-FLU (10% Ambersorb)	0	0
Control (10% Ambersorb)	100	0
CU-FLU (20% Zeolite)	0	0
Control (20% Zeolite)	88	18
CU-FLU (40% All Combination)	56	9
Control (40% All Combination)	72	33
CU-FLU (40% Control)	0	0
CU-FLU (20% Control)	0	0
CU-FLU (10% Control)	0	0
Control	96	9
Measured Copper	1040 mg/kg	
Measured Fluoranthene	121 mg/kg	

Table B-42. Solid-phase *H. azteca* TIE and Chemistry Results from Sediment Spiked with Copper and Fluoranthene (mean percent survival and standard deviation (SD)). CU-FLU Indicates Sediment Spiked with Copper and Fluoranthene.

	Test 2		SIR-300 Elution		Ambersorb Elution	
Treatment	Mean	SD	Mean	SD	Mean	SD
CU-FLU - Baseline	0	0				
CU-FLU (10% SIR-300)	6	9	0	0		
CU-FLU (10% Ambersorb)	42	24			0	0
CU-FLU (10% Control)	0	0				
Elution Control			27	31	100	0
Control	78	11	93	12	87	12
Measured Copper	804 mg/kg		53 mg/L			
Measured Fluoranthene	41.7 mg/kg				518 μg/L	

Table B-43. *H. azteca* TIE Results from Interstitial Water Extracted from Sediment Spiked with Copper and Fluoranthene (mean percent survival and standard deviation (SD) from Day 4). Interstitial Water Concentration of Copper is 18.8 mg/L and the Concentration of Fluoranthene is 179 µg/L.

	Percent Sample										
	0%		10%		25%		50%		100%		Toxic
Treatment	Mean	SD	Mean	SD	Mean	SD	Mean	SD	Mean	SD	Units
Baseline	100	0	0	0	0	0	0	0	0	0	≥20.0
Filtration	93	12	0	0	0	0	0	0	0	0	≥20.0
EDTA	87	12	60	40	13	23	0	0	0	0	7.2
Cation Rinsate	93	12	33	31	7	12	0	0	0	0	12.8
Cation Eluate	100	0	0	0	0	0	0	0	0	0	≥20.0
HLB Rinsate	73	12	0	0	0	0	0	0	0	0	≥20.0
HLB Eluate	83	29	60	20	7	12	0	0	0	0	7.2
Sequential Cation HLB	93	12	87	12	67	12	7	12	0	0	3.5
Sequential Cation Eluate	60	20	0	0	0	0	0	0	0	0	20.0
Sequential HLB Eluate	87	23	80	0	20	35	0	0	0	0	5.6

Table B-44. Extraction Efficiency Results for *H. azteca* TIE Treatments Conducted with Interstitial Water Extracted from Copper and Fluoranthene-spiked Sediment (mean percent survival and standard deviation (SD) from Day 2).

	Percent Sample										
	0%		25%		50%		100%		Toxic	[Cu]	[FLU]
Treatment	Mean	SD	Mean	SD	Mean	SD	Mean	SD	Units	mg/L	µg/L
Baseline	100	0	33	31	0	0	0	0	5.3	16.1	142
Cation Rinsate	100	0	33	42	47	12	0	0	4.6	1.69	ND
Cation Eluate	100	0	0	0	13	23	0	0	7.5	10.8	0.7
HLB Rinsate	100	0	40	0	0	0	0	0	4.8	13.0	ND
HLB Eluate	87	12	67	12	0	0	40	20	2.8	0.62	ND

Table B-45. Solid-phase *H. azteca* TIE Results from Sediment Spiked with Copper and Ammonia (mean percent survival and standard deviation (SD)). CU-AMM Indicates Sediment Spiked with Copper and Ammonia. Mean Total Ammonia Concentrations are from Sediment Overlying Water. The Combination Treatment Consists of SIR-300, Ambersorb, and Zeolite. Unionized Ammonia Concentrations are Calculated from Overlying Water Ammonia Concentrations, pH, Temperature, and Salinity.

Treatment	Test 1 Mean	Test 1 SD	Total [NH_3] mg/L	Unionized [NH_3] mg/L
CU-AMM - Baseline	0	0	127	5.70
CU-AMM (10% SIR-300)	0	0	72.9	7.87
Control (10% SIR-300)	98	4	8.5	1.68
CU-AMM (10% Ambersorb)	0	0	103.5	5.34
Control (10% Ambersorb)	100	0	10.9	1.27
CU-AMM (20% Zeolite)	0	0	24.8	2.57
Control (20% Zeolite)	100	0	19.0	1.39
CU-AMM (40% Combination)	62	11	18.5	3.07
Control (40% Combination)	100	0	2.7	0.64
CU-AMM (40% Control)	0	0	66.4	4.21
CU-AMM (20% Control)	0	0	94.2	5.29
CU-AMM (10% Control)	0	0	97.1	4.73
Control	96	5	12.1	1.72
Measured Copper	992 mg/kg			

Table B-46. Solid-phase *H. azteca* TIE and Chemistry Results from Sediment Spiked with Copper and Ammonia (mean percent survival and standard deviation (SD)). CU-AMM Indicates Sediment Spiked with Copper and Ammonia. Mean total Ammonia Concentrations are from Sediment Overlying Water and SIR-300 elution. Unionized Ammonia Concentrations are Calculated from Overlying Water Ammonia Concentrations, pH, Temperature, and Salinity.

Treatment	Test 2 Mean	Test 2 SD	Total [NH^3] mg/L	Unionized [NH^3] mg/L	SIR-300 Elution Mean	SIR-300 Elution SD
CU-AMM - Baseline	0	0	115	4.84		
CU-AMM (10% SIR-300)	0	0	57.6	6.04	0	0
CU-AMM (20% Zeolite)	0	0	33.4	2.79		
CU-AMM (20% Control)	0	0	76.4	3.56		
Elution Control					67	12
Control	100	0	0.30	0.05	87	12
Measured Copper	797 mg/kg				13.0 mg/L	

Table B-47. *H. azteca* TIE Results from Interstitial Water Extracted from Sediment Spiked with Copper and Ammonia (mean percent survival and standard deviation (SD) from Day 4). Interstitial Water Copper Concentration is 11.55 mg/L. Unionized Ammonia Concentrations are Calculated from Overlying Water Ammonia Concentrations, pH, Temperature, and Salinity.

	Percent Sample											Total	Unionized
	0%		10%		25%		50%		100%		Toxic	$[NH_3]$	$[NH_3]$
Treatment	Mean	SD	Mean	SD	Mean	SD	Mean	SD	Mean	SD	Units	mg/L	mg/L
Baseline	87	23	0	0	0	0	0	0	0	0	≥20	1560	6.09
Filtration	93	12	0	0	0	0	0	0	0	0	≥20	1520	11.3
Zeolite	73	23	0	0	0	0	0	0	0	0	≥20	760	5.38
Cation Column	100	0	73	12	0	0	0	0	0	0	7.5	1520	8.37
Sequential Zeolite EDTA	87	12	87	12	80	0	0	0	0	0	3.0	600	28.2
Seq. Zeolite Cation Rinsate	100	0	73	31	87	23	13	12	0	0	2.9	880	4.03
Seq. Zeolite Cation Eluate	87	12	89	19	47	12	0	0	0	0	4.1	NA	NA
Seq. Air Strip EDTA	73	12	81	2	80	20	0	0	0	0	2.8	890	30.0
Seq. Air Strip Cation Rinsate	100	0	81	20	60	20	0	0	0	0	4.0	880	22.7
Seq. Air Strip Cation Eluate	100	0	0	0	0	0	7	12	0	0	19.6	NA	NA

Table B-48. Extraction Efficiency Results for *H. azteca* TIE Treatments Conducted with Interstitial Water Extracted from Copper and Ammonia-spiked Sediment (mean percent survival and standard deviation (SD) from Day 4). Unionized Ammonia Concentrations are Calculated from Overlying Water Ammonia Concentrations, pH, Temperature, and Salinity.

	Percent Sample										Total	Unionized
	0%		25%		50%		100%		Toxic	[Cu]	$[NH_3]$	$[NH_3]$
Treatment	Mean	SD	Mean	SD	Mean	SD	Mean	SD	Units	mg/L	mg/L	mg/L
Baseline	93	12	0	0	0	0	0	0	8.0	11.0	1140	5.86
Cation Rinsate	81	20	0	0	0	0	0	0	8.0	0.17	1080	17.0
Cation Eluate	100	0	0	0	0	0	0	0	8.0	9.7	ND	ND
Zeolite Rinsate	93	12	0	0	0	0	0	0	8.0	4.4	410	7.72

Table B-49. Solid-phase *H. azteca* TIE Results from Sediment Spiked with Nonylphenol and Ammonia with (mean percent survival and standard deviation (SD)). NON-AMM Indicates Sediment Spiked with Nonylphenol and Ammonia. Total Ammonia Concentrations are from Sediment Overlying Water. The Combination Treatment Consists of SIR-300, Ambersorb, and Zeolite. Unionized Ammonia Concentrations are Calculated from Overlying Water Ammonia Concentrations, pH, Temperature, and Salinity.

	Test 1		Total $[NH_3]$	Unionized $[NH_3]$
Treatment	Mean	SD	mg/L	mg/L
NON-AMM - Baseline	0	0	143	4.96
NON-AMM (10% SIR-300)	0	0	79	5.71
Control (10% SIR-300)	100	0	11.1	1.20
NON-AMM (10% Ambersorb)	0	0	100	4.99
Control (10% Ambersorb)	96	9	12.1	0.87
NON-AMM (20% Zeolite)	0	0	33.5	2.13
Control (20% Zeolite)	100	0	3.90	0.24
NON-AMM (40% Combination)	48	26	23.4	3.18
Control (40% Combination)	100	0	2.40	0.36
NON-AMM (40% Control)	0	0	73.9	4.50
NON-AMM (20% Control)	0	0	105	5.66
NON-AMM (10% Control)	0	0	84.0	4.89
Control	100	0	15.2	1.20
Measured Nonylphenol	74 mg/kg			

Table B-50. Solid-phase *H. azteca* TIE and Chemistry Results from Sediment Spiked with Copper and Ammonia (mean percent survival and standard deviation (SD)). NON-AMM Indicates Sediment Spiked with Nonylphenol and Ammonia. Total Ammonia Concentrations are from Sediment Overlying Water and Ambersorb Elution. Unionized Ammonia Concentrations are Calculated from Overlying Water Ammonia Concentrations, pH, Temperature, and Salinity.

	Test 2		Total $[NH^3]$	Unionized $[NH^3]$	Ambersorb Elution	
Treatment	Mean	SD	mg/L	mg/L	Mean	SD
NON-AMM	0	0	115	3.24		
NON-AMM (10% Ambersorb)	14	15	90.6	3.34	0	0
NON-AMM (20% Zeolite)	0	0	32.6	1.87		
NON-AMM (20% Control)	0	0	45.9	1.76		
Elution Control					100	0
Control	100	0	0.10	0	87	12
Measured Nonylphenol	525 mg/kg				62 mg/L	

Table B-51. ***H. azteca*** **TIE Results from Interstitial Water Extracted from Sediment Spiked with Nonylphenol and Ammonia (mean percent survival and standard deviation (SD) from Day 7). Concentration of Nonylphenol in Interstitial Water is 714 µg/L. Unionized Ammonia Concentrations are Calculated from Overlying Water Ammonia Concentrations, pH, Temperature, and Salinity.**

	Percent Sample											Total	Unionized
	0%		10%		25%		50%		100%		Toxic	[NH_3]	[NH_3]
Treatment	Mean	SD	Mean	SD	Mean	SD	Mean	SD	Mean	SD	Units	mg/L	mg/L
Baseline	100	0	40	40	0	0	0	0	0	0	12.0	860	6.67
Filter	93	12	39	22	0	0	0	0	0	0	11.4	575	6.26
Zeolite Rinsate	87	12	87	12	93	12	47	12	0	0	1.9	218	2.86
HLB Rinsate	100	0	60	0	0	0	0	0	0	0	8.6	1020	15.7
Sequential Zeolite HLB Rinsate	100	0	100	0	100	0	76	8	0	0	1.6	386	7.60
Seq. Zeolite HLB Eluate	80	20	60	20	20	0	35	22	0	0	5.3	NA	NA
Seq. Air Strip HLB Rinsate	100	0	100	0	93	12	73	12	0	0	1.9	658	13.0
Seq. Air Strip HLB Eluate	100	0	80	35	80	20	53	31	27	12	1.9	NA	NA

Table B-52. Extraction Efficiency Results for ***H. azteca*** **TIE Treatments Conducted with Interstitial Water Extracted from Nonylphenol and Ammonia-spiked Sediment (mean percent survival and standard deviation (SD) from Day 4). Unionized Ammonia Concentrations are Calculated from Overlying Water Ammonia Concentrations, pH, Temperature, and Salinity.**

	Percent Sample										Total	Unionized
	0%		25%		50%		100%		Toxic	[NON]	[NH_3]	[NH_3]
Treatment	Mean	SD	Mean	SD	Mean	SD	Mean	SD	Units	µg/L	mg/L	mg/L
Baseline	80	20	0	0	0	0	0	0	8.0	8630	1050	10.0
HLB Rinsate	87	12	20	20	0	0	0	0	6.1	76.9	410	4.37
HLB Eluate	100	0	0	0	0	0	0	0	8.0	1370	ND	ND
Zeolite Rinsate	93	12	0	0	0	0	0	0	8.0	812	370	3.86

Table B-53. Initial Results from Switzer Creek Solid-phase and Interstitial Water Tests with *E. estuarius.* Mean Percent Survival and Standard Deviation (SD).

	Sediment		Interstitial Water		Sediment	
Treatment	Mean	SD	Mean	SD	Mean	SD
Control	100	0	80	42	100	0
5%					80	20
10%			60	52	40	20
15%					27	23
20%					0	0
25%	0	0	90	32		
50%	7	12	90	32		
75%	7	12				
100%	7	12	20	42		

Table B-54. Solid-phase Results from Switzer Creek TIEs with *E. estuarius.* Mean Percent Survival and Standard Deviation (SD).

	TIE 1 (25% SZ)		Ambersorb Elution		TIE 2 (25% SZ)		Ambersorb Elution		TIE 3 (10% SZ)	
Treatment	Mean	SD	Mean	SD	Mean	SD	Mean	SD	Mean	SD
SZ - Baseline	12	27			0	0			46	9
SZ (10% Ambersorb)	80	14							96	5
SZ (15% Ambersorb)	100	0	0	0	96	9	0	0		
Control (15% Ambersorb)	90	12	100	0	88	11	90	32		
SZ (20% SIR-300)	80	14	100	0	44	9	90	32		
SZ (10% SIR-300)									68	8
Control (20% SIR-300)	88	11	100	0	100	0	100	0		
SZ (20% Control)	24	17			4	9				
Control	100	0			96	9			98	2

Table B-55. Selected Switzer Creek Chemical Concentrations in Sediment and Related Extracts. Shaded Areas Indicate LC50 or Guideline was Exceeded. The Complete List of Undiluted Sediment Concentrations are Presented in Appendix C.

	100% Sediment	25% Sediment	Sediment LC50 or Guideline	Ambersorb Eluate 27 (μg/L)	Water LC50 (μg/L)	Tissue (ng/g)
Metal ug/g						
Copper	89.6	22.4	584 [1]	NA	NA	
Pyrethroid ng/g						
Bifenthrin	23.9	5.98	7.9 [1]	ND		ND
Bifenthrin ug/g oc	0.9	0.225	1.05 [1]	ND		ND
Cyfluthrin	183	45.7	13.7 [2]	ND		ND
Cyfluthrin ug/g oc	6.8	1.7	1.08 [2]	ND		ND
Permethrin	135	33.8	140 [1]	ND		ND
Permethrin ug/g oc	5	1.25	18.7 [1]	ND		ND
OC ng/g						
Oxadiazon	137	34.3		ND		28.7
Total Chlordane	62.6	15.7	>49 NOEC [3]	0.085	>310 NOEC [3]	
PAH ng/g						
Total PAH	6756	1689	10750 [4]	ND		NA
Total PAH ug/g oc	139	35	1800 [5]	ND		NA
PCB ng/g						
Total PCBs	90.4	22.6	400 MEC [6]	NA		ND
OP ng/g	ND	ND		NA		ND

[1] MPSL unpublished data, [2] Amweg et al., 2005, [3] Chris Stransky, Nautilus Environmental, San Diego, CA, U.S. , personal communication, [4] Page et al., 2002, LC50 for *Rhepoxynius abronius,* [5] Swartz 1999, Consensus Based Guideline Value, [6] MacDonald et al., 2000, Consensus Based Guideline Value

Table B-56. Initial Results from Consolidated Slip Solid-phase and Interstitial Water Tests with *E. estuarius.* Mean Percent Survival and Standard Deviation (SD).

	Sediment		Sediment		Interstitial Water	
Sample Concentration	Mean	SD	Mean	SD	Mean	SD
Control	100	0	100	0	100	0
1%			0	0	60	55
2%			0	0	20	45
5%			0	0		
10%	0	0	0	0	0	0
25%	0	0			0	0
50%	0	0				
75%	0	0				
100%	0	0				

Table B-57. Solid-phase Results from Consolidated Slip TIEs with *E. estuarius*. Mean Percent Survival and Standard Deviation (SD).

	TIE 1 (1% CS)		Ambersorb Elution		SIR-300 Elution	
Treatment	Mean	SD	Mean	SD	Mean	SD
CS - Baseline	12	11				
CS (10% SIR-300)	68	18			100	0
Control (10% SIR-300)	92	11			80	45
CS (10% Ambersorb)	100	0	0	0		
Control (10% Ambersorb)	100	0	100	0		
CS (20% SIR-300/Ambersorb)	100	0				
Control (20% SIR-300/Ambersorb)	100	0				
CS (20% Control)	48	23				
CS (10% Control)	44	41				
Control	100	0	100	0	80	45

Table B-58. *E. estuarius* TIE Results from First Test With Consolidated Slip Interstitial Water. Mean Percent Survival (10 day) and Standard Deviation (SD).

	Percent Sample									
	0%		1%		2%		5%		10%	
Treatment	Mean	SD	Mean	SD	Mean	SD	Mean	SD	Mean	SD
Baseline	100	0	0	0	0	0	0	0	0	0
Filtration	100	0	100	0	100	0	100	0	100	0
Aeration	100	0	0	0	0	0	0	0	0	0
EDTA	100	0	0	0	0	0	0	0	0	0
Cation Rinsate	100	0	100	0	100	0	100	0	80	45
Cation Eluate	100	0	100	0	80	45	100	0	100	0
HLB Rinsate	100	0	80	45	80	45	80	45	100	0
HLB Eluate	80	45	100	0	100	0	80	45	0	0
Sequential Cation HLB Rinsate	100	0	100	0	100	0	100	0	60	55
Sequential Cation Eluate	100	0	100	0	100	0	100	0	100	0
Sequential HLB Eluate	80	45	100	0	100	0	80	45	20	45

Table B-59. *E. estuarius* TIE Results from Second Test with Consolidated Slip Interstitial Water. Mean Percent Survival (10 day) and Standard Deviation (SD).

	Percent Sample										
	0%		0.25%		0.5%		1%		2%		Toxic
Treatment	Mean	SD	Mean	SD	Mean	SD	Mean	SD	Mean	SD	Units
Baseline	100	0	20	45	0	0	0	0	0	0	641
Centrifugation	100	0	20	45	0	0	0	0	0	0	641
Cation Rinsate	100	0	100	0	80	45	80	45	100	0	<50
Cation Eluate	100	0	100	0	100	0	80	45	100	0	<50
HLB Rinsate	100	0	100	0	100	0	80	45	100	0	<50
HLB Eluate	100	0	60	55	0	0	0	0	0	0	342
Sequential Cation HLB Rinsate	100	0	60	55	60	55	100	0	80	45	<50
Sequential Cation Eluate	60	55	80	45	100	0	100	0	100	0	<50
Sequential HLB Eluate	100	0	80	45	100	0	60	55	80	45	<50

Table B-60. *E. estuarius* TIE Results from Third Test with Consolidated Slip Interstitial Water. Mean Percent Survival (10 day) and Standard Deviation (SD). Column 2 Eluates are Percent Methanol.

	Percent Sample										
	0%		0.25%		0.5%		1%		2%		Toxic
Treatment	Mean	SD	Mean	SD	Mean	SD	Mean	SD	Mean	SD	Units
Baseline	100	0	20	45	40	55	0	0	0	0	667
Centrifugation	100	0	100	0	40	55	0	0	20	45	214
HLB Column 1 Rinsate	100	0	100	0	100	0	80	45	100	0	<50
HLB Column 1 Eluate	100	0	100	0	100	0	100	0	0	0	71
HLB Column 2 Rinsate	100	0	100	0	100	0	100	0	100	0	<50
HLB Column 2 Eluate 25%	100	0	100	0	100	0	100	0	100	0	<50
HLB Column 2 Eluate 50%	80	45	100	0	100	0	100	0	100	0	<50
HLB Column 2 Eluate 75%	80	45	80	45	100	0	80	45	100	0	<50
HLB Column 2 Eluate 80%	100	0	80	45	100	0	100	0	100	0	<50
HLB Column 2 Eluate 85%	100	0	100	0	80	45	60	55	100	0	<50
HLB Column 2 Eluate 90%	100	0	80	45	100	0	80	45	80	45	<50
HLB Column 2 Eluate 95%	100	0	80	45	100	0	100	0	60	55	<50
HLB Column 2 Eluate 100%	100	0	100	0	100	0	100	0	20	45	65

Table B-61. Selected Consolidated Slip Chemical Concentrations in Sediment, Interstitial Water, and Related Extracts. Shaded Areas Indicate LC50 or Guideline Was Exceeded. The Complete List of Undiluted Sediment and Interstitial Water Concentrations are Presented in Appendix C.

	100% Sediment	1% Sediment	Sediment LC50 or Guideline	Ambersorb Eluate (µg/L)	10% Interstitial Water (µg/L)	2% Interstitial Water (µg/L)	HLB Eluate (µg/L)	Water LC50 (µg/L)
Metal ug/g								
Antimony	16.7	0.167	2.5 ERM	NA	6.69	1.34	NA	
Chromium	767	7.67	370 ERM	NA	550	110	NA	
Copper	1040	10.4	584 [1]	NA	624	125	NA	33300 [1]
Lead	551	5.51	218 ERM	NA	564	113	NA	
Nickel	82.9	0.829	51.6 ERM	NA	44.9	8.98	NA	
Silver	8.79	0.0879	3.7 ERM	NA	ND	ND	NA	
Zinc	918	9.18	410 ERM	NA	637	127	NA	
Pyrethroid ng/g								
Bifenthrin	ND	ND	7.9	0.088	ND	ND	ND	
Cyfluthrin	ND	ND		0.250	ND	ND	ND	
Cypermethrin	ND	ND	11.4	0.180	ND	ND	ND	
(Es)fenvalerate	ND	ND		0.134	ND	ND	ND	
Permethrin	ND	ND	140	0.139	ND	ND	ND	
L-cyhalothrin	ND	ND		0.073	ND	ND	ND	
OC ng/g								
DDE(p,p')	115	1.15	27 ERM	ND	0.87	0.174	ND	
Dieldrin	9.25	0.09	8 ERM	ND	ND	ND	ND	
Total Chlordane	36.9	0.37	>49 NOEC[2]	ND	ND	ND	ND	
Total DDT	266	2.66	46.1 ERM	ND	1.36	0.272	ND	2.88 [3]
Total DDT ug/g oc	5.02	0.050	2500 [3]	ND	NA	NA	ND	
PAH ng/g								
Acenaphthene	2300	23	500 ERM	<RL	7	1.4	<RL	708 [4]
Phenanthrene	24300	243	1500 ERM	<RL	29	5.8	4.9	158 [4]
Fluoranthene	3360	33.6	85300 [1]	<RL	22	4.4	<RL	>70 [5]
Total PAH	1657822	16578	10750 [6]	NA	NA	NA	NA	
Total PAH ug/g oc	1126	11.3	1800 [7]	NA	NA	NA	NA	
PCB ng/g								
Total PCBs	349.4	3.5	400 MEC [8]	ND	ND	ND	ND	
OP ng/g	ND	ND		ND	ND	ND	ND	

[1] MPSL unpublished data, [2] Chris Stransky, Nautilus Environmental, San Diego, CA, U.S. , personal communication, [3] Swartz et al., 1994, [4] Swartz et al., 1995, [5] Boese et al., 1997, [6] Page et al., 2002, LC50 for *Rhepoxynius abronius,* [7] Swartz 1999, Consensus Based Guideline Value, [8] MacDonald et al., 2000a, Consensus Based Guideline Value

Table B-62. Initial Results from Upper Newport Solid-phase and Interstitial Water Tests with *E. estuarius.* Mean Percent Survival and Standard Deviation (SD).

	Sediment		Sediment		Interstitial Water	
Sample Concentration	Mean	SD	Mean	SD	Mean	SD
Control	100	0	100	0	100	0
5%			64	17		
10%			4	9	0	0
25%			0	0	0	0
50%			0	0	0	0
100%	0	0			0	0

Table B-63. Solid-phase Results from Upper Newport TIEs with *E. estuarius.* Mean Percent Survival and Standard Deviation (SD).

	TIE 1 (100% UN)		TIE 2 (10% UN)		Ambersorb Elution		TIE 3 (10% UN)	
Treatment	Mean	SD	Mean	SD	Mean	SD	Mean	SD
UN - Baseline	4	9	0	0			4	9
UN (10% SIR-300)	0	0	8	18				
Control (10% SIR-300)	100	0	100	0				
UN (10% Ambersorb)	0	0	12	18	0	0		
Control (10% Ambersorb)	96	9	100	0	100	0		
UN (10% PCC)							100	0
Control (10% PCC)							100	0
UN (20% SIR-300/Ambersorb)	0	0	72	11				
Control (20% SIR-300/Ambersorb)	88	18	92	18				
UN (20% Control)	0	0	4	9				
UN (10% Control)	0	0	0	0			4	9
UN (Enzyme)	0	0	4	9				
Control (Enzyme)	100	0	100	0				
UN (PBO)	0	0	4	9				
Control (PBO)	100	0	96	9				
Control	100	0	100	0	80	45	100	0

Table B-64. TIE Results from First Test with Upper Newport Interstitial Water. Mean Percent Survival (7 day) and Standard Deviation (SD).

	Percent Sample										
	0%		10%		25%		50%		100%		Toxic
Treatment	Mean	SD	Mean	SD	Mean	SD	Mean	SD	Mean	SD	Units
Baseline	80	45	100	0	0	0	0	0	20	45	6.3
Filtration	100	0	80	45	80	45	80	45	40	55	1.2
Aeration	100	0	100	0	60	55	0	0	0	0	3.9
EDTA	100	0	100	0	0	0	0	0	0	0	6.3
Cation Rinsate	100	0	100	0	100	0	40	55	20	45	2.1
Cation Eluate	100	0	100	0	100	0	100	0	100	0	<1
HLB Rinsate	100	0	100	0	60	55	40	55	20	45	2.7
HLB Eluate	80	45	100	0	100	0	100	0	100	0	<1
Sequential Cation HLB Rinsate	100	0	80	45	80	45	60	55	0	0	1.9
Sequential Cation Eluate	100	0	100	0	100	0	80	45	40	55	1.2
Sequential HLB Eluate	100	0	60	55	60	55	80	45	60	55	<1
Enzyme	80	45	100	0	100	0	80	45	20	45	1.4
PBO	100	0	0	0	0	0	0	0	0	0	20.0

Table B-65. TIE Results from Second Test with Upper Newport Interstitial Water. Mean Percent Survival (7 day) and Standard Deviation (SD).

	Percent Sample										
	0%		10%		25%		50%		100%		Toxic
Treatment	Mean	SD	Mean	SD	Mean	SD	Mean	SD	Mean	SD	Units
Baseline	100	0	100	0	60	55	0	0	0	0	3.9
Centrifugation	100	0	100	0	100	0	80	45	20	45	1.4
Cation Rinsate	100	0	100	0	80	45	100	0	0	0	1.6
Cation Eluate	100	0	100	0	60	55	80	45	100	0	<1.0
HLB Rinsate	100	0	100	0	100	0	40	55	0	0	2.1
HLB Eluate	100	0	100	0	100	0	100	0	100	0	<1.0
Sequential Cation HLB Rinsate	100	0	100	0	80	45	100	0	40	55	1.1
Sequential Cation Eluate	100	0	100	0	80	45	100	0	100	0	<1.0
Sequential HLB Eluate	100	0	80	45	100	0	80	45	0	0	1.6
Enzyme	100	0	100	0	100	0	60	55	60	55	<1.0
PBO	100	0	60	55	0	0	0	0	0	0	8.6

Table B-66. TIE Results from Second Test with Upper Newport Interstitial Water. Mean Percent Survival (10 day) and Standard Deviation (SD).

	Percent Sample										
	0%		10%		25%		50%		100%		Toxic
Treatment	Mean	SD	Mean	SD	Mean	SD	Mean	SD	Mean	SD	Units
Baseline	80	45	60	55	0	0	0	0	0	0	7.4
HLB Rinsate	80	45	100	0	60	55	20	45	0	0	3.2
HLB Eluate	80	45	60	55	100	0	100	0	80	45	<1
Enzyme	80	45	100	0	80	45	0	0	0	0	3.1
BSA	100	0	40	55	0	0	0	0	0	0	12.0
PBO	60	55	0	0	0	0	0	0	0	0	20.0

Table B-67. Selected Upper Newport Chemical Concentrations in Sediment, Interstitial Water, and Related Extracts. Shaded Areas Indicate LC50 or Guideline Was Exceeded. The Complete List of Undiluted Sediment and Interstitial Water Concentrations are Presented in Appendix C.

Chemical	100% Sediment	10% Sediment	Sediment LC50 or Guideline	Ambersorb Eluate 27 (µg/L)	Ambersorb Eluate 29 (µg/L)	100% Interstitial Water (µg/L)	Water LC50 (µg/L)
Metal ug/g							
Copper	32.2	0.32	584 [1]	NA	NA	3.87	33000 [1]
Silver	14.4	0.14	3.7 ERM	NA	NA	ND	
Pyrethroid ng/g							
Bifenthrin	28.1	2.81	7.90 [1]	0.028	0.018	ND	
Bifenthrin ug/g oc	1.8	0.180	1.05 [1]	NA	NA	ND	
(Es)fenvalerate	ND	ND		0.039	0.0043	ND	
L-cyhalothrin	ND	ND		0.047	0.038	ND	
Permethrin	6.72	0.672	140 [1]	0.077	0.017	ND	
Permethrin ug/g oc	0.4	0.040	18.7 [1]	NA	NA	ND	
OC ng/g							
Total Chlordane	18.4	0.184	>49 NOEC [2]	ND	ND	ND	
Total DDT	30.1	0.30	46.1 ERM	ND	ND	0.0210	2.88 [3]
Total DDT ug/g oc	1.89	0.019	2500 [3]	ND	ND	NA	
PAH ng/g							
Total PAH	2027	203	10750 [4]	NA	NA	NA	
Total PAH ug/g oc	27.7	2.77	1800 [5]	NA	NA	NA	
PCB ng/g							
Total PCBs	9.04	0.904	400 MEC [6]	ND	ND	ND	
OP ng/g	ND	ND		ND	ND	ND	

[1] MPSL unpublished data, [2] Chris Stransky, Nautilus Environmental, San Diego, CA, U.S. , personal communication, [3] Swartz et al., 1994, [4] Page et al., 2002, LC50 for *Rhepoxynius abronius,* [5] Swartz 1999, Consensus Based Guideline Value, [6] MacDonald et al., 2000a, Consensus Based Guideline Value

Table B-68. Initial Results from San Diego Creek (SD) Solid-phase and Interstitial Water Tests with *H. azteca*. Mean Percent Survival and Standard Deviation (SD).

	Sediment		Interstitial Water	
Sample Concentration	Mean	SD	Mean	SD
Control	96	5	93	12
10%			100	0
25%			100	0
50%			100	0
100%	16	23	100	0

Table B-69. Solid-phase Results from San Diego Creek (SD) TIEs with *H. azteca*. Mean Percent Survival and Standard Deviation (SD).

	TIE 1 (100% SD)		Ambersorb Elution		TIE 2 (100% SD)	
Treatment	Mean	SD	Mean	SD	Mean	SD
SD - Baseline	32	13			36	18
SD (10% SIR-300)	10	7				
Control (10% SIR-300)	100	0				
SD (10% Ambersorb)	86	11	0	0		
Control (10% Ambersorb)	100	0	93	12		
SD (20% SIR-300/Ambersorb)	0	0				
Control (20% SIR-300/Ambersorb)	94	9				
SD (20% Control)	98	4				
SD (10% Control)	94	9				
SD (Enzyme)	94	5			96	9
Control (Enzyme)	90	10			100	0
SD (BSA)					86	9
Control (BSA)					92	11
SD (PBO)	0	0				
Control (PBO)	98	4				
Control	90	14	93	12	94	9
15°C SD	2	4				
15°C SD (Enzyme)	84	15				
15°C Control (Enzyme)	98	4				
15°C Control	98	4				

Table B-70. Selected San Diego Creek Chemical Concentrations in Sediment and Related Extracts. Shaded Areas Indicate LC50 or Guideline Was Exceeded. The Complete List of Undiluted Sediment and Interstitial Water Concentrations are Presented in Appendix C.

Chemical	Test Sediment	Sediment LC50 or Guideline	Ambersorb Eluate 26 (µg/L)	Ambersorb Eluate 27 (µg/L)	Ambersorb Eluate 28 (µg/L)	Ambersorb Eluate 29 (µg/L)	Water LC50 (µg/L)
Metal ug/g							
Copper	2.22	260 [1]	NA	NA	NA	NA	
Pyrethroid ng/g							
Bifenthrin	ND		ND	0.025	0.019	0.078	0.0093 [2]
(Es)fenvalerate	ND		0.008	0.067	0.013	ND	
L-cyhalothrin	ND		0.010	0.031	0.037	ND	
Permethrin	ND		0.055	0.106	0.013	ND	0.0211 [2]
OC ng/g							
Total DDT	2.54	572 PEC	ND	ND	ND	ND	
Total DDT ug/g oc	0.348	371 [3]	NA	NA	NA	NA	
PAH ng/g							
Total PAH	23.1	22800 PEC	NA	NA	NA	NA	
Total PAH ug/g oc	3.17	1800 [4]					
PCB ng/g							
Total PCB	2.45	400 MEC [5]	ND	ND	ND	ND	
OP ng/g							
Dioxathion	NA		94	ND	ND	ND	8.6 [6]

[1] MPSL unpublished data, [2] Anderson et al., 2006, [3] Nebeker et al., 1989, [4] Swartz 1999, Consensus Based Guideline Value, [5] MacDonald et al., 2000a, Consensus Based Guideline Value, [6] Mayer and Ellersiack 1986, LC50 for *Gammarus fasciatus*

Table B-71. Initial Results from Indiana Harbor Solid-phase and Interstitial Water Tests with *H. azteca*. Mean Percent Survival and Standard Deviation (SD).

	Sediment		Sediment		Interstitial Water	
Sample Concentration	Mean	SD	Mean	SD	Mean	SD
Control	98	4	96	5	100	0
5%			90	14		
10%			88	11	24	21
25%			6	13	0	0
50%			6	9	0	0
100%	0	0			0	0

Table B-72. Solid-phase Results from Indiana Harbor (IH) TIEs with *H. azteca*. Mean Percent Survival and Standard Deviation (SD).

	TIE 1 (25% IH)		Ambersorb Elution		TIE 2 (25% IH)	
Treatment	Mean	SD	Mean	SD	Mean	SD
IH - Baseline	36	26			22	29
IH (10% SIR-300)	2	4				
Control (10% SIR-300)	98	4				
IH (10% Ambersorb)	14	15	0	0		
Control (10% Ambersorb)	96	5	87	12		
IH (10% PCC)					40	25
Control (10% PCC)					84	18
IH (10% Zeolite)	2	4				
Control (10% Zeolite)	100	0				
IH (30% SIR-300/Ambersorb/Zeolite)	22	13				
Control (30% SIR-300/Ambersorb/Zeolite)	98	4				
IH (30% Control)	30	16				
IH (10% Control)	0	0			6	5
IH (Enzyme)	84	11			78	13
Control (Enzyme)	96	5			100	0
IH (BSA)					86	17
Control (BSA)					92	11
Control	90	10	93	12	94	9
15°C IH	38	30				
15°C IH (Enzyme)	68	15				
15°C Control (Enzyme)	96	9				
15°C Control	90	10				

Table B-73. TIE Results from First Test With Indiana Harbor Interstitial Water. Only Ammonia Treatments Were Used to Determine if there Were Other Contaminants Contributing to Toxicity. Mean Percent Survival (7 day) and Standard Deviation (SD). TA Indicates Total Ammonia, UA Indicates Unionized Ammonia.

	Percent Sample									
	0%		5%		10%		20%		50%	
Treatment	Mean	SD	Mean	SD	Mean	SD	Mean	SD	Mean	SD
Baseline	80	20	0	0	0	0	0	0	0	0
	TA	UA	TA	UA	TA	UA	TA	UA	TA	UA
	0	0.000	17.9	1.383	36.4	2.475	91	4.567	182	6.278
Zeolite Rinsate	87	12	87	12	73	12	33	12	7	12
	TA	UA	TA	UA	TA	UA	TA	UA	TA	UA
	0	0.000	0	0.000	0	0.000	0	0.000	0.1	0.005
Air Strip	0	0	0	0	0	0	0	0	0	0
	TA	UA	TA	UA	TA	UA	TA	UA	TA	UA
	0.2	0.014	4.4	0.177	7.2	0.338	17.9	0.754	47.3	1.905

Table B-74. TIE Results from Second Test with Indiana Harbor Interstitial Water. Ammonia was Removed with a Zeolite Column Prior to TIE. Mean Percent Survival (7 day) and Standard Deviation (SD).

	Percent Sample										
	0%		5%		10%		20%		50%		Toxic
Treatment	Mean	SD	Mean	SD	Mean	SD	Mean	SD	Mean	SD	Units
Baseline	100	0	93	12	100	0	93	12	0	0	2.9
Aeration	100	0	100	0	93	12	60	20	7	12	3.9
EDTA	87	23	100	0	81	2	33	58	7	12	5.1
Cation Rinsate	87	12	93	12	74	13	100	0	80	20	<1
Cation Eluate	93	12	100	0	100	0	100	0	93	12	<1
HLB Rinsate	87	12	93	12	87	12	100	0	100	0	<1
HLB Eluate	93	12	87	12	100	0	93	12	100	0	<1
Sequential Cation HLB Rinsate	100	0	100	0	93	12	93	12	80	20	<1
Sequential Cation El	87	12	93	12	100	0	93	12	100	0	<1
Sequential HLB El	93	12	100	0	93	12	93	12	93	12	<1
Enzyme	87	12	87	12	87	12	93	12	33	31	2.3
BSA	93	12	100	0	100	0	88	11	27	23	2.6
PBO	53	42	100	0	60	0	0	0	0	0	7.5

Table B-75. TIE Results from Third Test with Indiana Harbor Interstitial Water. Ammonia Was Removed with a Zeolite Column After TIE Treatments. Mean Percent Survival (7 day) and Standard Deviation (SD).

	Percent Sample										
	0%		5%		10%		20%		50%		Toxic
Treatment	Mean	SD	Mean	SD	Mean	SD	Mean	SD	Mean	SD	Units
Baseline	59	34	53	12	33	31	0	0	0	0	4.1
Cation Rinsate	77	9	60	17	40	20	27	31	0	0	3.7
Cation Eluate	84	15	80	20	80	20	67	23	56	14	<1
HLB Column	73	31	67	12	93	12	60	20	7	12	1.6
HLB Methanol Eluate	87	12	93	12	93	12	0	0	0	0	2.8
HLB Acetone Eluate	80	20	67	23	33	23	7	12	0	0	5.0

Table B-76. Selected Indiana Harbor Chemical Concentrations in Sediment, Interstitial Water, and Related Extracts. Shaded Areas Indicate LC50 or Guideline Was Exceeded. The Complete List of Undiluted Sediment and Interstitial Water Concentrations are Presented in Appendix C.

Chemical	100% Sediment	25% Sediment	Sediment LC50 or Guideline	Ambersorb Eluate (µg/L)	100% Interstitial Water (µg/L)	50% Interstitial Water (µg/L)	HLB Eluate (µg/L)	Water LC50 (µg/L)
Metal ug/g								
Arsenic	37	9.25	33 PEC	NA	40.2	20.1	NA	
Cadmium	9.39	2.35	4.98 PEC	NA	2.07	1.04	NA	2.8 [2]
Chromium	867	217	111 PEC	NA	484	242	NA	
Copper	336	84.0	260 [1]	NA	262	131	NA	31 [2]
Lead	786	197	128 PEC	NA	944	472	NA	<16 [2]
Nickel	93	23.3	48.6 PEC	NA	101	50.5	NA	780 [2]
Zinc	3910	978	459 PEC	NA	1730	865	NA	73 [2]
Pyrethroid ng/g	ND	ND		ND	ND	ND	ND	
OC ng/g								
DDE(p,p')	ND	ND		ND	0.465	0.233	ND	1.66 [4]
Endosulfan I	ND	ND		ND	2.35	1.18	ND	
Total Chlordane	37.5	9.37	17.6 PEC	ND	ND	ND	ND	
Total DDT	154	38.4	572 PEC	ND	0.465	0.233	ND	
Total DDT ug/g oc	0.764	0.191	371 [3]	ND	NA	NA	ND	
PAH ng/g								
Anthracene	4600	1150	845 PEC	0.0026	19	9.5	<RL	5.6 [5]
Fluoranthene	35300	8825	90600 [1]	0.0499	214	107	<RL	44 [6]
Total PAH	236850	59213	22800 PEC	NA	NA	NA	NA	
Total PAH ug/g oc	1030	907	1800 [7]	NA	NA	NA	NA	
PCB ng/g								
Total PCBs	7365	1841	1700 EEC [8]	ND	9.87	4.93	ND	
Total PCBs (aroclor)	NA	NA		ND	19.7	9.87	ND	56.4 LOEC [9]
OP ng/g	ND	ND		ND	ND	ND	ND	

[1] MPSL unpublished data, [2] Phipps et al., 1995, [3] Nebeker et al., 1989, [4] Hoke et al., 1994, [5] Hatch and Burton, 1999, [6] Suedel et al., (1993), [7] Swartz 1999, Consensus Based Guideline Value, [8] MacDonald et al., 2000a, Consensus Based Guideline Value, [9] Borgmann et al., 1990

Table B-77. Initial Results from Alisal Creek Solid-phase and Interstitial Water Tests with *H. azteca*. Mean Percent Survival and Standard Deviation (SD).

Sample Concentration	Sediment		Interstitial Water	
	Mean	SD	Mean	SD
Control	96	5	100	0
10%	100	0	63	15
25%	84	11	13	23
50%	40	22	0	0
100%	0	0	0	0

Table B-78. Solid-phase Results from Alisal Creek (AC) TIEs with *H. azteca*. Mean Percent Survival and Standard Deviation (SD).

Treatment	TIE 1 (100% AC)		Ambersorb Elution		SIR-300 Elution	
	Mean	SD	Mean	SD	Mean	SD
AC - Baseline	0	0				
AC (10% SIR-300)	0	0			80	20
Control (10% SIR-300)	96	5			100	0
AC (10% Ambersorb)	0	0	0	0		
Control (10% Ambersorb)	86	5	47	31		
AC (20% SIR-300/Ambersorb)	0	0				
Control (20% SIR-300/Ambersorb)	70	35				
AC (20% Control)	0	0				
AC (10% Control)	0	0				
AC (Enzyme)	72	16				
Control (Enzyme)	76	43				
AC (BSA)	0	0				
Control (BSA)	90	7				
AC (PBO)	0	0				
Control (PBO)	84	5				
Control	94	5	67	58	100	0
15°C AC	0	0				
15°C AC (Enzyme)	0	0				
15°C Control (Enzyme)	88	8				
15°C Control	98	4				

Table B-79. TIE Results from First Test With Alisal Creek Interstitial Water. Mean Percent Survival (10 day) and Standard Deviation (SD).

	Percent Sample										
	0%		5%		10%		20%		50%		Toxic
Treatment	Mean*	SD*	Mean	SD	Mean	SD	Mean	SD	Mean	SD	Units
Baseline	93	12	60	0	13	23	0	0	0	0	7.7
Aeration	93	12	40	35	0	0	0	0	0	0	11.4
EDTA	93	12	0	0	0	0	0	0	0	0	20.0
Cation Rinsate	87	23	93	12	81	2	33	23	20	20	2.2
Cation Eluate	87	12	73	23	93	12	87	23	73	23	<1
HLB Rinsate	93	12	100	0	93	12	60	20	0	0	1.9
HLB Eluate	93	12	87	12	73	31	0	0	0	0	3.3
Sequential Cation HLB Rinsate	100	0	93	12	100	0	76	21	20	20	1.5
Sequential Cation Eluate	67	23	80	0	73	12	87	23	93	12	<1
Sequential HLB Eluate	87	23	100	0	73	31	80	20	67	12	<1
Enzyme	93	12	87	12	73	23	0	0	0	0	3.3
BSA	73	12	0	0	0	0	0	0	0	0	20.0
PBO	82	17	0	0	0	0	0	0	0	0	20.0

Table B-80. TIE Results from Second Test With Alisal Creek Interstitial Water. Mean Percent Survival (10 day) and Standard Deviation (SD).

	Percent Sample										
	0%		5%		10%		20%		50%		Toxic
Treatment	Mean	SD	Mean	SD	Mean	SD	Mean	SD	Mean	SD	Units
Baseline	87	12	60	0	0	0	0	0	0	0	7.8
HLB Rinsate	93	12	88	11	83	29	80	20	60	20	<1
HLB Methanol Eluate	93	12	73	12	67	12	0	0	0	0	3.4
HLB Acetone Eluate	80	0	20	20	0	0	0	0	0	0	14.9

Table B-81. Selected Alisal Creek Chemical Concentrations in Sediment, Interstitial Water, and Related Extracts. Shaded Areas Indicate LC50 or Guideline was Exceeded. The Complete List of Undiluted Sediment and Interstitial Water Concentrations are Presented in Appendix C.

Chemical	Test Sediment	Sediment LC50 or Guideline	Ambersorb Eluate (µg/L)	100% Interstitial Water (µg/L)	50% Interstitial Water (µg/L)	HLB Eluate (µg/L)	Water LC50 (µg/L)
Metal ug/g							
Copper	29.2	260 [1]	NA	68.8	34.4	NA	31 [2]
Lead	16.8	128 PEC	NA	65.7	32.9	NA	<16 [2]
Pyrethroid ng/g							
Bifenthrin	ND	ND	0.104	ND	ND	ND	0.0093 [4]
Cyfluthrin	ND	ND	0.429	ND	ND	ND	
Cypermethrin	ND	ND	0.242	ND	ND	ND	0.0053 [10]
(Es)fenvalerate	6.93	41.8 [3]	0.138	ND	ND	ND	
(Es)fenvalerate ug/g oc	0.4	1.54 [3]	NA	NA	NA	NA	
L-cyhalothrin	26.77	5.6 [3]	0.129	0.051	0.025	ND	
L-cyhalothrin ug/g oc	1.4	0.45 [3]	NA	NA	NA	NA	
Permethrin	ND	201 [3]	0.888	0.603	0.302	ND	0.0211 [4]
OC ng/g							
DDD(o,p')	175		0.005	0.117	0.058	ND	
DDD(p,p')	1588	28 PEC	ND	0.277	0.138	0.006	0.180 [6]
DDE(p,p')	444	31.3 PEC	0.004	0.785	0.393	ND	1.66 [6]
DDT(p,p')	537	62.9 PEC	ND	0.070	0.035	ND	0.07 [6]
Total Chlordane	21.6	17.6 PEC	ND	ND	ND	ND	
Total DDT	2753	572 PEC	NA	1.25	0.624	NA	
Total DDT ug/g oc	143	371 [5]	NA	NA	NA	NA	
PAH ng/g							
Total PAH	787	22800 PEC	NA	NA	NA	NA	
Total PAH ug/g oc	41	1800 - 7	NA	NA	NA	NA	
PCB ng/g							
Total PCBs	101.4	400 MEC [8]	ND	ND	ND	ND	
OP ng/g							
Chlorpyrifos	ND		0.379	ND	ND	ND	0.086 [2]
Diazinon	ND		0.440	ND	ND	ND	6.51 [9]

[1] MPSL unpublished data, [2] Phipps et al., 1995, [3] Amweg et al., 2005, [4] Anderson et al., 2006, [5] Nebeker et al., 1989, [6] Hoke et al., 1994, [7] Swartz 1999, Consensus Based Guideline Value, [8] MacDonald et al., 2000a, Consensus Based Guideline Value, [9] Ankley and Collyard, 1995, [10] Solomon et al., 2001

APPENDIX C

CHEMISTRY DATA TABLES

Table C-1. Concentrations of Selected Compounds in Formulated Sediments (µg/L).

Analyte	Formulated 1	Formulated 2	Formulated 3
Organochlorine Pesticides			
Aldrin	ND	ND	ND
Chlordane, cis	ND	ND	ND
Chlordane, trans	ND	ND	ND
Chlordene, alpha	ND	ND	ND
Chlordene, gamma	ND	ND	ND
Dacthal	8.16	2.99	ND
DDD, o,p'	ND	ND	ND
DDD, p,p'	ND	ND	ND
DDE, o,p'	ND	ND	ND
DDE, p,p'	2.28	5.34	1.73
DDMU, p,p'	ND	ND	ND
DDT, o,p'	ND	ND	ND
DDT, p,p'	ND	ND	ND
Dieldrin	ND	ND	ND
Endosulfan I	ND	ND	ND
Endosulfan II	ND	ND	ND
Endosulfan sulfate	ND	ND	ND
Endrin	ND	ND	ND
Endrin Aldehyde	ND	ND	ND
Endrin Ketone	ND	ND	ND
HCH, alpha	ND	ND	ND
HCH, beta	ND	ND	ND
HCH, gamma	ND	ND	ND
Heptachlor	ND	ND	ND
Heptachlor epoxide	ND	ND	ND
Hexachlorobenzene	ND	ND	ND
Methoxychlor	ND	ND	ND
Mirex	ND	ND	ND
Nonachlor, cis	ND	ND	ND
Nonachlor, trans	ND	ND	ND
Oxadiazone	ND	ND	ND
Oxychlordane	ND	ND	ND
Tedion	ND	ND	ND
Organophosphate Pesticides			
Chlorpyrifos	ND	ND	ND
Chlorpyrifos methyl	ND	ND	ND
Diazinon	ND	ND	ND
Dichlofenthion	ND	ND	ND
Dioxathion	ND	ND	ND
Ethion	ND	ND	ND
Fenchlorophos (Ronnel)	ND	ND	ND
Fonofos (Dyfonate)	ND	ND	ND
Parathion	ND	ND	ND

Analyte	Formulated 1	Formulated 2	Formulated 3
Sulfotepp	ND	ND	ND
Thionzin(Thionazin)	ND	ND	ND
Tokuthion	ND	ND	ND
Tributylphosphorotrithioite (Merphos)	ND	ND	ND
Trichloronate	ND	ND	ND
Fenitrothion	ND	ND	ND
Malathion	ND	ND	ND
Ethoprop (Prophos)	ND	ND	ND
Pyrethroid Pesticides			
Bifenthrin	ND	ND	ND
Cyfluthrin-1	ND	ND	ND
Cyfluthrin-2	ND	ND	ND
Cyfluthrin-3	ND	ND	ND
Cyfluthrin-4	ND	ND	ND
Cypermethrin-1	ND	ND	ND
Cypermethrin-2	ND	ND	ND
Cypermethrin-3	ND	ND	ND
Cypermethrin-4	ND	ND	ND
Deltamethrin	ND	ND	ND
Esfenvalerate/Fenvalerate-1	ND	ND	ND
Esfenvalerate/Fenvalerate-2	ND	ND	ND
Permethrin-1	ND	ND	ND
Permethrin-2	ND	ND	ND
Warrior-1	NA	NA	ND
Warrior-2	NA	NA	ND
Physical Parameters			
TOC	0.55	0.78	0.91
% Cobble	0.00	0.00	0.00
%Gravel	0.00	0.00	0.00
%Sand - Coarse	0.00	0.00	0.00
%Sand - Medium	18.71	13.57	6.77
%Sand - Fine	49.97	48.17	46.35
%Fines - Silt	7.02	13.49	11.29
%Fines - Clay	24.3	24.78	35.59

Table C-2. Concentrations of Trace Metals in Sediment (µg/kg).

Metals	MDL	RL	Switzer Creek	Consolidated Slip	Upper Newport	San Diego Creek	Indiana Harbor	Alisal Creek
Aluminum (Al)	1	5	17700	27200	29900	2870	11700	24900
Antimony (Sb)	0.025	0.05	1.20	16.7	1.09	0.110	10.8	0.720
Arsenic (As)	0.025	0.05	5.44	25.2	7.15	1.06	37	10.7
Barium (Ba)	0.025	0.05	90.2	1430	143	40.9	211	247
Beryllium (Be)	0.025	0.05	0.270	0.580	0.620	0.080	0.920	0.620
Cadmium (Cd)	0.025	0.05	1.32	6.19	2.16	0.510	9.39	1.37
Chromium (Cr)	0.025	0.05	32.1	767	34.3	3.25	867	53.9
Cobalt (Co)	0.025	0.05	5.96	15	7.83	1.05	8.44	11.8
Copper (Cu)	0.025	0.05	89.6	1040	32.2	2.22	336	29.2
Iron (Fe)	1	5	22000	41800	28300	3780	144000	32400
Lead (Pb)	0.025	0.05	80.5	551	13.7	1.78	786	16.8
Manganese (Mn)	0.025	0.05	153	502	306	94.2	1640	457
Mercury (Hg)	0.00001	0.00002	0.250	3.82	0.0292	0.0028	0.0973	0.033
Molybdenum (Mo)	0.025	0.05	4.26	5.54	2.89	0.510	19.6	1.43
Nickel (Ni)	0.025	0.05	15.9	82.9	18.4	1.99	93	42.5
Selenium (Se)	0.025	0.05	0.270	3.65	1.24	0.220	6.40	0.520
Silver (Ag)	0.025	0.05	1.56	8.79	14.4	4.24	8.94	6.89
Strontium (Sr)	0.025	0.05	49.9	352	112	26.7	62.4	51.7
Thallium (Tl)	0.025	0.05	0.220	0.340	0.370	0.050	2.77	0.450
Tin (Sn)	0.025	0.05	7.52	13.2	2.95	1.23	843	5.40
Titanium (Ti)	0.025	0.05	1200	13.2	1630	481	312	1850
Vanadium (V)	0.025	0.05	48.3	80.7	81	10.8	35.1	63.9
Zinc (Zn)	0.025	0.05	303	918	143	12.8	3910	140

Table C-3. Concentrations of Pyrethroid Pesticides in Sediment (µg/kg).

Pyrethroid Pesticides	MDL	RL	Switzer Creek	Consolidated Slip	Upper Newport	San Diego Creek	Indiana Harbor	Alisal Creek
Bifenthrin	0.500	1.00	23.9	ND	28.1	ND	ND	ND
Cyfluthrin-1	1.50	3.00	52.7	ND	ND	ND	ND	ND
Cyfluthrin-2	1.50	3.00	58.0	ND	ND	ND	ND	ND
Cyfluthrin-3	1.50	3.00	49.2	ND	ND	ND	ND	ND
Cyfluthrin-4	1.50	3.00	22.9	ND	ND	ND	ND	ND
Cypermethrin-1	1.50	3.00	ND	ND	ND	ND	ND	ND
Cypermethrin-2	1.50	3.00	ND	ND	ND	ND	ND	ND
Cypermethrin-3	1.50	3.00	ND	ND	ND	ND	ND	ND
Cypermethrin-4	1.50	3.00	ND	ND	ND	ND	ND	ND
Esfenvalerate/Fenvalerate-1	1.00	2.00	ND	ND	ND	ND	ND	ND
Esfenvalerate/Fenvalerate-2	1.00	2.00	ND	ND	ND	ND	ND	6.93
Lambda-cyhalothrin-1	0.500	1.00	ND	ND	ND	ND	ND	4.70
Lambda-cyhalothrin-2	0.500	1.00	ND	ND	ND	ND	ND	22.1
Permethrin-1	2.00	4.00	15.2	ND	6.72	ND	ND	ND
Permethrin-2	2.00	4.00	120	ND	ND	ND	ND	ND
Deltamethrin	0.500	1.00	ND	ND	ND	ND	ND	ND
Total Cyfluthrin	NA	NA	182.7	NA	NA	NA	NA	NA
Total Lambda-cyhalothrin	NA	NA	NA	NA	NA	NA	NA	26.77
Total Permethrin	NA	NA	135	NA	6.72	NA	NA	NA
Bifenthrin µg/g oc	NA	NA	0.9	NA	1.1	NA	NA	NA
Cyfluthrin µg/g oc	NA	NA	5.0	NA	NA	NA	NA	NA
Esfenvalerate µg/g oc	NA	NA	NA	NA	NA	NA	NA	0.3
Lambda-cyhalothrin µg/g oc	NA	NA	NA	NA	NA	NA	NA	1.0
Permethrin µg/g oc	NA	NA	6.8	NA	0.3	NA	NA	NA

Table C-4. Concentrations of Organochlorine Pesticides in Marine Sediment (µg/kg).

Organochlorine Pesticides	MDL	RL	Switzer Creek	MDL	RL	Consolidated Slip	MDL	RL	Upper Newport
Aldrin	0.465	1.79	0	0.473	1.82	10.2	0.509	1.96	0
Chlordane, cis-	1.28	1.79	18.8	1.30	1.82	11.8	1.40	1.96	5.90
Chlordane, trans-	0.722	1.79	18.6	0.735	1.82	9.63	0.791	1.96	7.51
Chlordene, alpha-	0.493	0.894	2.74	0.502	0.910	6.32	0.540	0.979	2.10
Chlordene, gamma-	0.458	0.894	1.23	0.466	0.910	0	0.501	0.979	1.48
Dacthal	1.13	1.79	0	1.15	1.82	0	1.24	1.96	0
DDD(o,p')	1.37	1.79	2.80	1.40	1.82	36.5	1.50	1.96	0
DDD(p,p')	1.61	1.79	7.65	1.64	1.82	106	1.76	1.96	5.02
DDE(o,p')	1.20	3.58	0	1.22	3.64	8.92	1.32	3.92	0
DDE(p,p')	1.03	3.58	13.3	1.05	3.64	115	1.13	3.92	25.1
DDMU(p,p')	2.15	5.36	0	2.19	5.46	17.2	2.36	5.87	0
DDT(o,p')	1.82	5.36	0	1.85	5.46	0	1.99	5.87	0
DDT(p,p')	4.42	8.94	6.79	4.50	9.10	0	4.84	9.79	0
Dieldrin	0.751	0.894	4.18	0.764	0.910	9.25	0.822	0.979	1.18
Endosulfan I	1.93	3.58	0	1.97	3.64	0	2.11	3.92	0
Endosulfan II	7.15	17.9	0	7.28	18.2	0	7.83	19.6	0
Endosulfan sulfate	7.15	17.9	0	7.28	18.2	0	7.83	19.6	0
Endrin	1.68	3.58	0	1.71	3.64	0	1.84	3.92	0
HCH, alpha	0.851	0.894	0	0.866	0.910	0	0.932	0.979	0
HCH, beta	1.10	1.79	0	1.12	1.82	0	1.21	1.96	0
HCH, gamma	0.608	0.894	0	0.619	0.910	0	0.666	0.979	0
Heptachlor	0.922	1.79	0	0.939	1.82	0	1.01	1.96	0
Heptachlor epoxide	0.901	1.79	0	0.917	1.82	0	0.987	1.96	0
Hexachlorobenzene	0.193	0.536	0	0.197	0.546	0	0.211	0.587	0
Methoxychlor	2.65	5.36	0	2.69	5.46	0	2.90	5.87	0
Mirex	1.69	2.68	0	1.72	2.73	3.81	1.85	2.94	0
Nonachlor, cis-	1.75	1.79	7.30	1.78	1.82	0	1.92	1.96	0
Nonachlor, trans-	0.694	1.79	17.9	0.706	1.82	15.5	0.760	1.96	4.96
Oxadiazon	1.67	1.79	137	1.70	1.82	0	1.83	1.96	38.3
Oxychlordane	0.658	1.79	0	0.67	1.82	0	0.720	1.96	0
Tedion	1.32	3.58	0	1.34	3.64	0	1.44	3.92	0
Toxaphene	14.3	35.8	0	14.6	36.4	0	15.7	39.2	0
Total Chlordane			62.6			36.9			18.4
Total DDT			30.5			266			30.1
Total DDT oc			1.14			5.02			1.89

Table C-5. Concentrations of Organochlorine Pesticides in Freshwater Sediment (µg/kg).

Organochlorine Pesticides	MDL	RL	San Diego Creek	MDL	RL	Indiana Harbor	MDL	RL	Alisal Creek
Aldrin	0.294	1.13	0	0.593	2.28	349	0.421	1.62	3.94
Chlordane, cis-	0.811	1.13	0	1.63	2.28	20.6	1.16	1.62	6.02
Chlordane, trans-	0.458	1.13	0	0.922	2.28	12.8	0.654	1.62	7.41
Chlordene, alpha-	0.313	0.566	0	0.630	1.14	22.6	0.447	0.810	1.65
Chlordene, gamma-	0.290	0.566	0	0.584	1.14	56.2	0.415	0.810	0
Dacthal	0.716	1.13	0	1.44	2.28	0	1.02	1.62	12.6
DDD(o,p')	0.870	1.13	0	1.75	2.28	0	1.24	1.62	175
DDD(p,p')	1.02	1.13	0	2.05	2.28	6.40	1.46	1.62	1588
DDE(o,p')	0.761	2.27	0	1.53	4.57	136	1.09	3.24	3.98
DDE(p,p')	0.652	2.27	2.54	1.31	4.57	0	0.933	3.24	444
DDMU(p,p')	1.36	3.40	0	2.75	6.85	0	1.95	4.86	35.8
DDT(o,p')	1.15	3.40	0	2.32	6.85	11.2	1.65	4.86	5.17
DDT(p,p')	2.80	5.66	0	5.64	11.4	0	4.00	8.10	537
Dieldrin	0.476	0.566	0	0.959	1.14	7.50	0.680	0.810	52.8
Endosulfan I	1.22	2.27	0	2.47	4.57	0	1.75	3.24	2.43
Endosulfan II	4.53	11.3	0	9.13	22.8	0	6.48	16.2	8.92
Endosulfan sulfate	4.53	11.3	0	9.13	22.8	0	6.48	16.2	9.08
Endrin	1.06	2.27	0	2.15	4.57	0	1.52	3.24	11.9
HCH, alpha	0.539	0.566	0	1.09	1.14	0	0.771	0.810	0
HCH, beta	0.698	1.13	0	1.41	2.28	0	0.998	1.62	0
HCH, gamma	0.385	0.566	0	0.776	1.14	0	0.551	0.810	0
Heptachlor	0.584	1.13	0	1.18	2.28	243	0.836	1.62	0
Heptachlor epoxide	0.571	1.13	0	1.15	2.28	0	0.816	1.62	0
Hexachlorobenzene	0.122	0.340	0	0.247	0.685	0	0.175	0.486	2.76
Methoxychlor	1.68	3.40	0	3.38	6.85	0	2.40	4.86	0
Mirex	1.07	1.70	0	2.15	3.42	0	1.53	2.43	0
Nonachlor, cis-	1.11	1.13	0	2.24	2.28	4.08	1.59	1.62	0
Nonachlor, trans-	0.439	1.13	0	0.886	2.28	0	0.629	1.62	8.14
Oxadiazon	1.06	1.13	2.54	2.14	2.28	0	1.52	1.62	10.3
Oxychlordane	0.417	1.13	0	0.840	2.28	0	0.596	1.62	0
Tedion	0.834	2.27	0	1.68	4.57	0	1.19	3.24	4.21
Toxaphene	9.06	22.7	0	18.3	45.7	0	13.0	32.4	0
Total Chlordane			0			37.5			21.6
Total DDT			2.54			154			2753
Total DDT oc			0.35			0.76			143

Table C-6. Concentrations of Polycyclic Aromatic Hydrocarbons in Sediment (µg/kg).

Polycyclic Aromatic Hydrocarbons	MDL	RL	Switzer Creek	Consolidated Slip	Upper Newport	San Diego Creek	Indiana Harbor	Alisal Creek
Naphthalene	0.0025	0.005	43.9	1940	7.16	<RL	4650	6.49
Methylnaphthalene, 2-	0.0025	0.005	27	527	4.36	<RL	6940	7.74
Methylnaphthalene, 1-	0.0025	0.005	15.2	15500	8.94	<RL	3660	4.02
Dimethylnaphthalene, 2,6-	0.0025	0.005	32.8	25100	32.9	<RL	13100	12.6
Trimethylnaphthalene, 2,3,5-	0.0025	0.005	5.76	18000	12.3	<RL	3880	4.7
Naphthalenes, C1 -	0.0025	0.005	43	17600	14.5	<RL	10700	12.2
Naphthalenes, C2 -	0.0025	0.005	83.9	133000	105	2.89	45800	45.2
Naphthalenes, C3 -	0.0025	0.005	99	251000	148	5.18	78000	103
Naphthalenes, C4 -	0.0025	0.005	35.9	76400	77.2	2.55	18900	37.6
Biphenyl	0.0025	0.005	13.2	1550	<RL	<RL	995	2.35
Acenaphthylene	0.0025	0.005	38.8	892	2.27	<RL	2280	2.27
Acenaphthene	0.0025	0.005	18.2	2300	2.16	<RL	6620	6.86
Fluorene	0.0025	0.005	30.6	5890	5.6	<RL	5680	10.8
Methylfluorene, 1-	0.0025	0.005	11.3	18100	14.7	<RL	4380	6.91
Fluorenes, C1 -	0.0025	0.005	33	38900	37.2	<RL	13600	19.7
Fluorenes, C2 -	0.0025	0.005	33.8	38700	60.1	<RL	18200	34.7
Fluorenes, C3 -	0.0025	0.005	68.7	19400	47.3	<RL	11300	52.2
Dibenzothiophene	0.0025	0.005	23	5250	6.37	<RL	2560	6.93
Methyldibenzothiophene, 4-	0.0025	0.005	18.1	11000	11.5	<RL	7740	12.9
Dibenzothiophenes, C1 -	0.0025	0.005	44.3	22800	27.6	<RL	15300	27.2
Dibenzothiophenes, C2 -	0.0025	0.005	136	396300	73.6	<RL	28200	80.8
Dibenzothiophenes, C3 -	0.0025	0.005	166	19900	73.9	<RL	23800	76.7
Phenanthrene	0.0025	0.005	256	24300	46.6	2.17	17600	64.1
Methylphenanthrene, 1-	0.0025	0.005	40.6	23200	21.8	<RL	5000	13.1
Dimethylphenanthrene, 3,6-	0.0025	0.005	24.4	17600	20.9	<RL	5340	13.9
Phenanthrene/Anthracene, C1 -	0.0025	0.005	211	130000	136	3.83	34000	87.6
Phenanthrene/Anthracene, C2 -	0.0025	0.005	354	223000	225	5.57	49800	147
Phenanthrene/Anthracene, C3 -	0.0025	0.005	259	132000	190	3.44	39800	115
Phenanthrene/Anthracene, C4 -	0.0025	0.005	73.6	15700	39.8	<RL	7900	13.8
Anthracene	0.0025	0.005	74.2	4650	5.66	<RL	4600	8.2
Fluoranthene	0.0025	0.005	677	3360	67	3.66	35300	156
Methylfluoranthene, 2-	0.0025	0.005	48.8	2630	7.58	<RL	3690	11
Fluoranthene/Pyrenes, C1 -	0.0025	0.005	522	36300	96.6	2.91	44500	120
Pyrene	0.0025	0.005	683	9900	77.8	3.77	35000	138
Benz[a]anthracene	0.0025	0.005	178	2510	19.8	1.85	18000	36
Chrysene	0.0025	0.005	257	3970	37.6	2.4	24200	63.5
Chrysenes, C1 -	0.0025	0.005	137	14200	46.6	1.63	14900	26.8
Chrysenes, C2 -	0.0025	0.005	111	14300	53.7	2.06	6300	16
Chrysenes, C3 -	0.0025	0.005	81.3	5790	27	<RL	2250	9.59
Benzo(b)fluoranthene	0.0025	0.005	394	916	54.2	3.05	24500	79.9
Benzo(k)fluoranthene	0.0025	0.005	131	222	19.4	<RL	9100	25.3
Benzo(e)pyrene	0.0025	0.005	250	1600	40.6	2.12	16000	44.5
Benzo(a)pyrene	0.0025	0.005	285	1480	33.3	2.54	19700	44
Perylene	0.0025	0.005	101	637	20.4	<RL	5310	13.9
Indeno(1,2,3-c,d)pyrene	0.0025	0.005	328	267	31.6	<RL	11700	44.8
Dibenz(a,h)anthracene	0.0025	0.005	80.9	264	10.1	<RL	3770	12.7
Benzo(g,h,i)perylene	0.0025	0.005	401	634	59.8	<RL	11000	83.4
Total PAH (Page et al., 2002)			6756.3	1657822	2026.52	51.62	721815	1875.09
Total LMW PAH			1108.1	37442	130.85	5.83	71050	243.92
Total HMW PAH			2605	22358	309.1	17.27	165800	542.7
Total PAH			3713.1	59800	439.95	23.1	236850	786.62
Total PAH oc			139.1	1126.2	27.7	3.2	1177.8	41.0

Table C-7. Concentrations of Polychlorinated Biphenyls in Marine Sediment (µg/kg).

Polychlorinated Biphenyls	MDL	RL	Switzer Creek	MDL	RL	Consolidated Slip	MDL	RL	Upper Newport
PCB 008	0.179	0.358	0.375	2.00	4.00	5.14	0.196	0.392	0.289
PCB 018	0.179	0.358	1.08	2.00	4.00	14.1	0.196	0.392	0.365
PCB 027	0.179	0.358	0	2.00	4.00	0	0.196	0.392	0
PCB 028	0.179	0.358	4.62	2.00	4.00	17.5	0.196	0.392	0.546
PCB 029	0.179	0.358	0	2.00	4.00	0	0.196	0.392	0
PCB 031	0.179	0.358	2.44	2.00	4.00	18.8	0.196	0.392	0.492
PCB 033	0.179	0.358	1.16	2.00	4.00	13.2	0.196	0.392	0.306
PCB 044	0.179	0.358	6.36	2.00	4.00	31.8	0.196	0.392	0.816
PCB 049	0.179	0.358	5.38	2.00	4.00	21.9	0.196	0.392	0.446
PCB 052	0.179	0.358	8.53	2.00	4.00	40.6	0.196	0.392	0.941
PCB 056	0.179	0.358	2.12	2.00	4.00	12.9	0.196	0.392	0
PCB 060	0.179	0.358	0.952	2.00	4.00	8.19	0.196	0.392	0
PCB 066	0.179	0.358	3.81	2.00	4.00	23.4	0.196	0.392	0.665
PCB 070	0.179	0.358	5.47	2.00	4.00	35.2	0.196	0.392	1.11
PCB 074	0.179	0.358	1.82	2.00	4.00	13.9	0.196	0.392	1.13
PCB 087	0.179	0.358	4.85	2.00	4.00	16	0.196	0.392	0.735
PCB 095	0.179	0.358	9.61	2.00	4.00	35.4	0.196	0.392	1.34
PCB 097	0.179	0.358	3.39	2.00	4.00	12.2	0.196	0.392	0.591
PCB 099	0.179	0.358	4.75	2.00	4.00	15.6	0.196	0.392	0.72
PCB 101	0.179	0.358	11.7	2.00	4.00	40.6	0.196	0.392	1.27
PCB 105	0.179	0.358	6.46	2.00	4.00	12.9	0.196	0.392	0.464
PCB 110	0.179	0.358	15.9	2.00	4.00	38.0	0.196	0.392	2.09
PCB 114	0.179	0.358	0	2.00	4.00	0	0.196	0.392	0
PCB 118	0.179	0.358	9.57	2.00	4.00	28.2	0.196	0.392	1.34
PCB 128	0.179	0.358	4.19	2.00	4.00	6.55	0.196	0.392	0
PCB 137	0.179	0.358	0.626	2.00	4.00	0	0.196	0.392	0
PCB 138	0.179	0.358	12.8	2.00	4.00	33.6	0.196	0.392	1.06
PCB 141	0.179	0.358	2.12	2.00	4.00	8.45	0.196	0.392	0
PCB 149	0.179	0.358	7.99	2.00	4.00	33	0.196	0.392	0.735
PCB 151	0.179	0.358	1.83	2.00	4.00	10.1	0.196	0.392	0
PCB 153	0.179	0.358	11.9	2.00	4.00	39.6	0.196	0.392	0.772
PCB 156	0.179	0.358	1.37	2.00	4.00	3.68	0.196	0.392	0
PCB 157	0.179	0.358	0.357	2.00	4.00	0	0.196	0.392	0
PCB 158	0.179	0.358	1.67	2.00	4.00	2.79	0.196	0.392	0
PCB 170	0.179	0.358	1.97	2.00	4.00	9.24	0.196	0.392	0
PCB 174	0.179	0.358	1.97	2.00	4.00	12.1	0.196	0.392	0
PCB 177	0.179	0.358	1.31	2.00	4.00	5.91	0.196	0.392	0
PCB 180	0.179	0.358	3.74	2.00	4.00	25.7	0.196	0.392	0.313
PCB 183	0.179	0.358	1.12	2.00	4.00	7.15	0.196	0.392	0
PCB 187	0.179	0.358	2.35	2.00	4.00	14.5	0.196	0.392	0
PCB 189	0.179	0.358	0	2.00	4.00	0	0.196	0.392	2.38
PCB 194	0.179	0.358	0.834	2.00	4.00	6.28	0.196	0.392	0
PCB 195	0.179	0.358	0	2.00	4.00	0	0.196	0.392	0
PCB 200	0.179	0.358	0	2.00	4.00	0	0.196	0.392	0
PCB 201	0.179	0.358	1.11	2.00	4.00	6.28	0.196	0.392	0
PCB 203	0.179	0.358	0.728	2.00	4.00	5.36	0.196	0.392	0
PCB 206	0.179	0.358	0.566	2.00	4.00	3.84	0.196	0.392	0
PCB 209	0.179	0.358	0.375	2.00	4.00	2.17	0.196	0.392	0.2
Aroclor 1248	18	45	88	200	500	490	20	49	0
Aroclor 1254	7.0	18	160	80	200	300	7.8	20	12
Aroclor 1260	7.0	18	29	80	200	180	7.8	20	0
Total PCB's			90.396			349.44			9.041

Table C-8. Concentrations of Polychlorinated Biphenyls in Freshwater Sediment (µg/kg).

Polychlorinated Biphenyls	MDL	RL	San Diego Creek	MDL	RL	Indiana Harbor	MDL	RL	Alisal Creek
PCB 008	0.113	0.227	0	2.51	5.03	286	0.162	0.324	0.364
PCB 018	0.113	0.227	0	23	46.2	983	0.162	0.324	1.12
PCB 027	0.113	0.227	0	2.51	5.03	58.9	0.162	0.324	0
PCB 028	0.113	0.227	0.179	23	46.2	1304	0.162	0.324	1.75
PCB 029	0.113	0.227	0	2.51	5.03	7.45	0.162	0.324	0
PCB 031	0.113	0.227	0	23	46.2	1193	0.162	0.324	1.62
PCB 033	0.113	0.227	0	2.51	5.03	577	0.162	0.324	0.955
PCB 044	0.113	0.227	0.22	23	46.2	1200	0.162	0.324	2.68
PCB 049	0.113	0.227	0	23	46.2	1179	0.162	0.324	2.11
PCB 052	0.113	0.227	0.224	23	46.2	1358	0.162	0.324	4.15
PCB 056	0.113	0.227	0	2.51	5.03	459	0.162	0.324	0.973
PCB 060	0.113	0.227	0	2.51	5.03	229	0.162	0.324	0.339
PCB 066	0.113	0.227	0	2.51	5.03	872	0.162	0.324	2.32
PCB 070	0.113	0.227	0.301	23	46.2	1417	0.162	0.324	3.53
PCB 074	0.113	0.227	0.32	2.51	5.03	495	0.162	0.324	0.515
PCB 087	0.113	0.227	0	2.51	5.03	188	0.162	0.324	4.26
PCB 095	0.113	0.227	0.299	2.51	5.03	335	0.162	0.324	8.61
PCB 097	0.113	0.227	0	2.51	5.03	160	0.162	0.324	2.96
PCB 099	0.113	0.227	0	2.51	5.03	204	0.162	0.324	4.05
PCB 101	0.113	0.227	0.375	2.51	5.03	375	0.162	0.324	11.6
PCB 105	0.113	0.227	0	2.51	5.03	168	0.162	0.324	4.19
PCB 110	0.113	0.227	0.541	2.51	5.03	387	0.162	0.324	12.6
PCB 114	0.113	0.227	0	2.51	5.03	13.2	0.162	0.324	0
PCB 118	0.113	0.227	0.405	2.51	5.03	304	0.162	0.324	8.9
PCB 128	0.113	0.227	0	2.51	5.03	28.9	0.162	0.324	3.21
PCB 137	0.113	0.227	0	2.51	5.03	9.96	0.162	0.324	0.67
PCB 138	0.113	0.227	0.346	2.51	5.03	132	0.162	0.324	17.1
PCB 141	0.113	0.227	0	2.51	5.03	37.4	0.162	0.324	4.06
PCB 149	0.113	0.227	0.245	2.51	5.03	135	0.162	0.324	15.2
PCB 151	0.113	0.227	0	2.51	5.03	40.4	0.162	0.324	4.8
PCB 153	0.113	0.227	0.279	2.51	5.03	158	0.162	0.324	18.5
PCB 156	0.113	0.227	0	2.51	5.03	18.2	0.162	0.324	1.69
PCB 157	0.113	0.227	0	2.51	5.03	3.97	0.162	0.324	0
PCB 158	0.113	0.227	0	2.51	5.03	0	0.162	0.324	1.23
PCB 170	0.113	0.227	0	2.51	5.03	37.3	0.162	0.324	4.76
PCB 174	0.113	0.227	0.123	2.51	5.03	46.9	0.162	0.324	6.23
PCB 177	0.113	0.227	0	2.51	5.03	25.5	0.162	0.324	3.5
PCB 180	0.113	0.227	0.116	2.51	5.03	94.3	0.162	0.324	12
PCB 183	0.113	0.227	0	2.51	5.03	24.8	0.162	0.324	2.85
PCB 187	0.113	0.227	0.117	2.51	5.03	47.3	0.162	0.324	6.89
PCB 189	0.113	0.227	1.56	2.51	5.03	1.09	0.162	0.324	0
PCB 194	0.113	0.227	0	2.51	5.03	17	0.162	0.324	1.93
PCB 195	0.113	0.227	0	2.51	5.03	7.46	0.162	0.324	0.925
PCB 200	0.113	0.227	0	2.51	5.03	2.27	0.162	0.324	0.279
PCB 201	0.113	0.227	0	2.51	5.03	22.4	0.162	0.324	2.43
PCB 203	0.113	0.227	0	2.51	5.03	21.5	0.162	0.324	1.38
PCB 206	0.113	0.227	0	2.51	5.03	7.16	0.162	0.324	0.432
PCB 209	0.113	0.227	0.189	2.51	5.03	2.35	0.162	0.324	0.511
Aroclor 1248	11	28	0	2300	5800	34000	16	40	50
Aroclor 1254	4.5	11	0	920	2300	<2300	6.5	16	140
Aroclor 1260	4.5	11	0	920	2300	0	6.5	16	76
Total PCB's			2.45			7364.77			101.402

Table C-9. Concentrations of Organophosphate Pesticides in Sediment (µg/kg).

Organophosphate Pesticides	MDL	RL	Switzer Creek	Consolidated Slip	Upper Newport	San Diego Creek	Indiana Harbor	Alisal Creek
Chlorpyrifos	5.00	10.0	ND	ND	ND	ND	ND	ND
Chlorpyrifos methyl	25.0	50.0	ND	ND	ND	ND	ND	ND
Diazinon	5.0	10.0	ND	ND	ND	ND	ND	ND
Dichlofenthion	25.0	50.0	ND	ND	ND	ND	ND	ND
Dioxathion	25.0	50.0	ND	ND	ND	ND	ND	ND
Ethion	25.0	50.0	ND	ND	ND	ND	ND	ND
Fenchlorphos	25.0	50.0	ND	ND	ND	ND	ND	ND
Fenitrothion	25.0	50.0	ND	ND	ND	ND	ND	ND
Fonofos	25.0	50.0	ND	ND	ND	ND	ND	ND
Malathion	25.0	50.0	ND	ND	ND	ND	ND	ND
Parathion, Methyl	10.0	20.0	ND	ND	ND	ND	ND	ND
Parathion, Ethyl	10.0	20.0	ND	ND	ND	ND	ND	ND
Phosphamidon	25.0	50.0	ND	ND	ND	ND	ND	ND
Ethoprop	25.0	50.0	ND	ND	ND	ND	ND	ND
Sulfotep	25.0	50.0	ND	ND	ND	ND	ND	ND
Thionazin	25.0	50.0	ND	ND	ND	ND	ND	ND
Tokuthion	25.0	50.0	ND	ND	ND	ND	ND	ND
Merphos	25.0	50.0	ND	ND	ND	ND	ND	ND
Trichloronate	10.0	20.0	ND	ND	ND	ND	ND	ND

Table C-10. Concentration of Total Organic Carbon (TOC) and Summary of Grain Size Parameters.

Metals	MDL	Switzer Creek	Consolidated Slip	Upper Newport	San Diego Creek	Indiana Harbor	Alisal Creek
TOC	0.01	2.67	5.31	1.59	0.73	20.11	1.92
% Cobble	NA	0.00	0.00	0.00	0.00	0.00	0.00
%Gravel	NA	0.00	0.00	0.00	0.00	0.00	0.29
%Sand - Coarse	NA	0.00	0.00	0.00	0.19	0.02	1.32
%Sand - Medium	NA	1.36	0.67	0.35	16.18	0.46	9.30
%Sand - Fine	NA	45.95	25.06	23.85	79.94	8.08	12.63
%Fines - Silt	NA	29.14	38.12	39.87	1.87	62.76	34.75
%Fines - Clay	NA	23.55	36.15	35.93	1.82	28.67	41.71

Table C-11. Concentrations of Trace Metals in Interstitial Water (µg/L).

Metals	MDL	RL	Consolidated Slip	Upper Newport	Indiana Harbor	Alisal Creek
Aluminum (Al)	1	5	5150	530	3710	5910
Antimony (Sb)	0.1	0.5	6.69	1.08	14.2	5.37
Arsenic (As)	0.1	0.5	29.9	30.9	40.2	50.4
Barium (Ba)	0.1	0.5	NA	NA	161	471
Beryllium (Be)	0.1	0.5	0.235	0.049	0.227	0.347
Cadmium (Cd)	0.1	0.2	4.88	0.233	2.07	0.496
Chromium (Cr)	0.1	0.5	550	3.86	484	16.1
Cobalt (Co)	0.1	0.5	3.95	0.667	ND	ND
Copper (Cu)	0.1	0.5	624	3.87	262	68.8
Iron (Fe)	1	5	6250	759	17100	20100
Lead (Pb)	0.05	0.1	564	1.98	944	65.7
Manganese (Mn)	0.1	0.5	135	1670	399	10200
Mercury (Hg)	0.0005	0.001	2.00	0.972	1.27	0.963
Molybdenum (Mo)	0.1	0.5	4.98	0.384	74.8	4.44
Nickel (Ni)	0.1	0.5	44.9	2.03	101	26.5
Selenium (Se)	0.1	0.5	1.91	ND	9.92	3.34
Silver (Ag)	0.1	0.2	ND	ND	ND	ND
Strontium (Sr)	0.5	1	NA	NA	239	716
Thallium (Tl)	0.1	0.5	0.114	ND	3.36	0.117
Tin (Sn)	0.1	0.5	1.85	0.080	162	ND
Titanium (Ti)	0.1	0.5	245	48.3	47.3	302
Vanadium (V)	0.1	0.5	29	9.70	9.36	30.6
Zinc (Zn)	0.1	0.5	637	6.12	1730	208

Table C-12. Concentrations of Pyrethroid Pesticides in Interstitial Water (µg/L).

Pyrethroid Pesticides	MDL	RL	Consolidated Slip	Upper Newport	Indiana Harbor	Alisal Creek
Bifenthrin	0.0005	0.005	ND	ND	ND	ND
Cyfluthrin-1	0.002	0.022	ND	ND	ND	ND
Cyfluthrin-2	0.002	0.022	ND	ND	ND	ND
Cyfluthrin-3	0.002	0.022	ND	ND	ND	ND
Cyfluthrin-4	0.002	0.022	ND	ND	ND	ND
Cypermethrin-1	0.002	0.022	ND	ND	ND	ND
Cypermethrin-2	0.002	0.022	ND	ND	ND	ND
Cypermethrin-3	0.002	0.022	ND	ND	ND	ND
Cypermethrin-4	0.002	0.022	ND	ND	ND	ND
Esfenvalerate/Fenvalerate-1	0.001	0.010	ND	ND	ND	ND
Esfenvalerate/Fenvalerate-2	0.001	0.010	ND	ND	ND	ND
Lambda-cyhalothrin-1	0.0005	0.005	ND	ND	ND	ND
Lambda-cyhalothrin-2	0.0005	0.005	ND	ND	ND	0.051
Permethrin-1	0.003	0.025	ND	ND	ND	0.426
Permethrin-2	0.003	0.025	ND	ND	ND	0.177
Deltamethrin	0.002	0.025	ND	ND	ND	ND

Table C-13. Concentrations of Organochlorine Pesticides in Interstitial Water (μg/L).

Organochlorine Pesticides	MDL	RL	Consolidated Slip	Upper Newport	Indiana Harbor	Alisal Creek
Aldrin	0.001	0.010	ND	ND	ND	0.026
Chlordane, cis-	0.001	0.010	ND	0.0072*	ND	ND
Chlordane, trans-	0.001	0.010	ND	ND	ND	ND
Chlordene, alpha-	0.001	0.010	ND	ND	ND	ND
Chlordene, gamma-	0.001	0.010	ND	ND	ND	ND
Dacthal	0.001	0.010	ND	ND	ND	ND
DDD(o,p')	0.001	0.010	ND	ND	ND	0.117
DDD(p,p')	0.001	0.010	0.870	ND	ND	0.277
DDE(o,p')	0.001	0.010	ND	0.00686*	ND	ND
DDE(p,p')	0.001	0.010	0.491	0.0138	0.465	0.785
DDMU(p,p')	0.001	0.010	ND	ND	ND	ND
DDT(o,p')	0.001	0.010	ND	ND	ND	ND
DDT(p,p')	0.002	0.025	ND	ND	ND	0.0698
Dieldrin	0.001	0.010	ND	ND	ND	ND
Endosulfan I	0.001	0.010	ND	ND	2.35	ND
Endosulfan II	0.001	0.010	ND	ND	ND	ND
Endosulfan sulfate	0.001	0.010	ND	ND	ND	ND
Endrin	0.001	0.010	ND	ND	ND	ND
Endrin Aldehyde	0.002	0.025	ND	ND	ND	ND
Endrin Ketone	0.002	0.025	ND	ND	ND	ND
HCH, alpha	0.001	0.010	ND	ND	ND	ND
HCH, beta	0.001	0.010	ND	ND	ND	ND
HCH, gamma	0.001	0.010	ND	ND	ND	ND
HCH, delta	0.001	0.010	ND	ND	ND	ND
Heptachlor	0.001	0.010	ND	ND	ND	ND
Heptachlor epoxide	0.001	0.010	ND	ND	ND	ND
Hexachlorobenzene	0.0005	0.005	ND	ND	ND	ND
Methoxychlor	0.001	0.010	ND	ND	ND	ND
Mirex	0.001	0.010	ND	ND	ND	ND
Nonachlor, cis-	0.001	0.010	ND	ND	ND	ND
Nonachlor, trans-	0.001	0.010	ND	ND	ND	ND
Oxadiazon	0.001	0.010	ND	ND	ND	0.0696
Oxychlordane	0.001	0.010	ND	ND	ND	ND
Tedion	0.001	0.010	ND	ND	ND	ND

Table C-14. Concentrations of Polycyclic Aromatic Hydrocarbons in Interstitial Water (µg/L).

Polycyclic Aromatic Hydrocarbons	MDL	RL	Consolidated Slip	Upper Newport	Indiana Harbor	Alisal Creek
Naphthalene	0.0025	0.005	<RL	<RL	<RL	<RL
Methylnaphthalene, 2-	0.0025	0.005	<RL	<RL	12	<RL
Methylnaphthalene, 1-	0.0025	0.005	<RL	<RL	16	<RL
Dimethylnaphthalene, 2,6-	0.0025	0.005	19	<RL	67	0.00689
Trimethylnaphthalene, 2,3,5-	0.0025	0.005	66	<RL	33	<RL
Naphthalenes, C1 -	0.0025	0.005	7	<RL	29	<RL
Naphthalenes, C2 -	0.0025	0.005	149	<RL	246	0.00856
Naphthalenes, C3 -	0.0025	0.005	254	<RL	168	<RL
Naphthalenes, C4 -	0.0025	0.005	617	<RL	164	<RL
Biphenyl	0.0025	0.005	<RL	<RL	<RL	<RL
Acenaphthylene	0.0025	0.005	<RL	<RL	8	<RL
Acenaphthene	0.0025	0.005	7	<RL	62	<RL
Fluorene	0.0025	0.005	9	<RL	18	<RL
Methylfluorene, 1-	0.0025	0.005	54	<RL	32	<RL
Fluorenes, C1 -	0.0025	0.005	141	<RL	91	<RL
Fluorenes, C2 -	0.0025	0.005	338	<RL	132	<RL
Fluorenes, C3 -	0.0025	0.005	393	<RL	161	<RL
Dibenzothiophene	0.0025	0.005	12	<RL	10	<RL
Methyldibenzothiophene, 4-	0.0025	0.005	26	<RL	54	<RL
Dibenzothiophenes, C1 -	0.0025	0.005	80	<RL	101	<RL
Dibenzothiophenes, C2 -	0.0025	0.005	200	<RL	206	0.01380
Dibenzothiophenes, C3 -	0.0025	0.005	178	0.0353	191	0.03020
Phenanthrene	0.0025	0.005	29	<RL	15	<RL
Methylphenanthrene, 1-	0.0025	0.005	55	<RL	40	<RL
Dimethylphenanthrene, 3,6-	0.0025	0.005	64	<RL	38	<RL
Phenanthrene/Anthracene, C1 -	0.0025	0.005	318	0.00866	202	<RL
Phenanthrene/Anthracene, C2 -	0.0025	0.005	2658	0.0227	505	<RL
Phenanthrene/Anthracene, C3 -	0.0025	0.005	1878	0.0376	181	<RL
Phenanthrene/Anthracene, C4 -	0.0025	0.005	326	0.015	75	<RL
Anthracene	0.0025	0.005	12	<RL	19	<RL
Fluoranthene	0.0025	0.005	22	0.0187	214	<RL
Methylfluoranthene, 2-	0.0025	0.005	22	<RL	25	<RL
Fluoranthene/Pyrenes, C1 -	0.0025	0.005	417	0.0238	310	<RL
Pyrene	0.0025	0.005	71	0.0202	201	<RL
Benz[a]anthracene	0.0025	0.005	10	<RL	80	<RL
Chrysene	0.0025	0.005	17	0.0356	83	<RL
Chrysenes, C1 -	0.0025	0.005	72	0.0206	64	<RL
Chrysenes, C2 -	0.0025	0.005	87	0.0196	38	0.01860
Chrysenes, C3 -	0.0025	0.005	40	<RL	12	<RL
Benzo(b)fluoranthene	0.0025	0.005	<RL	0.0228	71	<RL
Benzo(k)fluoranthene	0.0025	0.005	<RL	<RL	22	<RL
Benzo(e)pyrene	0.0025	0.005	10	0.0174	52	<RL
Benzo(a)pyrene	0.0025	0.005	6	<RL	60	<RL
Perylene	0.0025	0.005	<RL	<RL	8	<RL
Indeno(1,2,3-c,d)pyrene	0.0025	0.005	<RL	<RL	25	<RL
Dibenz(a,h)anthracene	0.0025	0.005	<RL	<RL	12	<RL
Benzo(g,h,i)perylene	0.0025	0.005	6	<RL	32	<RL

Table C-15. Concentrations of Polychlorinated Biphenyls in Interstitial Water (μg/L).

Polychlorinated Biphenyls	MDL	RL	Consolidated Slip	Upper Newport	Indiana Harbor	Alisal Creek
PCB 005	0.010	0.020	ND	ND	ND	ND
PCB 008	0.010	0.020	ND	ND	3.03	ND
PCB 015	0.010	0.020	ND	ND	ND	ND
PCB 018	0.010	0.020	ND	ND	3.90	ND
PCB 027	0.010	0.020	ND	ND	ND	ND
PCB 028	0.010	0.020	ND	ND	ND	ND
PCB 029	0.010	0.020	ND	ND	ND	ND
PCB 031	0.010	0.020	ND	ND	2.06	ND
PCB 033	0.010	0.020	ND	ND	1.95	ND
PCB 044	0.010	0.020	ND	ND	2.09	ND
PCB 049	0.010	0.020	ND	ND	2.30	ND
PCB 052	0.010	0.020	ND	ND	ND	ND
PCB 056	0.010	0.020	ND	ND	0.992	ND
PCB 060	0.010	0.020	ND	ND	0.720	ND
PCB 066	0.010	0.020	ND	ND	ND	ND
PCB 070	0.010	0.020	ND	ND	2.18	ND
PCB 074	0.010	0.020	ND	ND	1.14	ND
PCB 087	0.010	0.020	ND	ND	ND	ND
PCB 095	0.010	0.020	ND	ND	ND	ND
PCB 097	0.010	0.020	ND	ND	ND	ND
PCB 099	0.010	0.020	ND	ND	ND	ND
PCB 101	0.010	0.020	ND	ND	ND	ND
PCB 105	0.010	0.020	ND	ND	0.592	ND
PCB 110	0.010	0.020	ND	ND	1.20	ND
PCB 114	0.010	0.020	ND	ND	ND	ND
PCB 118	0.010	0.020	ND	ND	0.260	ND
PCB 128	0.010	0.020	ND	ND	ND	ND
PCB 137	0.010	0.020	ND	ND	ND	ND
PCB 138	0.010	0.020	ND	ND	ND	ND
PCB 141	0.010	0.020	ND	ND	ND	ND
PCB 149	0.010	0.020	ND	ND	ND	ND
PCB 151	0.010	0.020	ND	ND	ND	ND
PCB 153	0.010	0.020	ND	ND	ND	ND
PCB 156	0.010	0.020	ND	ND	ND	ND
PCB 157	0.010	0.020	ND	ND	ND	ND
PCB 158	0.010	0.020	ND	ND	ND	ND
PCB 170	0.010	0.020	ND	ND	ND	ND
PCB 174	0.010	0.020	ND	ND	ND	ND
PCB 177	0.010	0.020	ND	ND	ND	ND
PCB 180	0.010	0.020	ND	ND	ND	ND
PCB 183	0.010	0.020	ND	ND	ND	ND
PCB 187	0.010	0.020	ND	ND	ND	ND
PCB 189	0.010	0.020	ND	ND	ND	ND
PCB 194	0.010	0.020	ND	ND	ND	ND
PCB 195	0.010	0.020	ND	ND	ND	ND
PCB 200	0.010	0.020	ND	ND	ND	ND
PCB 201	0.010	0.020	ND	ND	ND	ND
PCB 203	0.010	0.020	ND	ND	ND	ND
PCB 206	0.010	0.020	ND	ND	ND	ND
PCB 209	0.010	0.020	ND	ND	ND	ND

Table C-16. Concentrations of Organophosphate Pesticides in Interstitial Water (µg/L).

Organophosphate Pesticides	MDL	RL	Consolidated Slip	Upper Newport	Indiana Harbor	Alisal Creek
Aspon	0.030	0.100	ND	ND	ND	ND
Azinphos ethyl	0.030	0.100	ND	ND	ND	ND
Carbophenothion	0.030	0.100	ND	ND	ND	ND
Chlorfenvinphos	0.030	0.100	ND	ND	ND	ND
Chlorpyrifos	0.020	0.100	ND	ND	ND	ND
Chlorpyrifos methyl	0.020	0.100	ND	ND	ND	ND
Ciodrin	0.030	0.100	ND	ND	ND	ND
Coumaphos	0.040	0.100	ND	ND	ND	ND
Demeton-s	0.040	0.100	ND	ND	ND	ND
Diazinon	0.005	0.040	ND	ND	ND	ND
Naled	0.030	0.100	ND	ND	ND	ND
Dichlofenthion	0.030	0.100	ND	ND	ND	ND
Dichlorvos	0.030	0.100	ND	ND	ND	ND
Dicrotophos	0.030	0.100	ND	ND	ND	ND
Dimethoate	0.030	0.100	ND	ND	ND	ND
Dioxathion	0.030	0.100	ND	ND	ND	ND
Disulfoton	0.010	0.100	ND	ND	ND	ND
Ethion	0.020	0.100	ND	ND	ND	ND
Famphur	0.030	0.100	ND	ND	ND	ND
Fenchlorphos	0.030	0.100	ND	ND	ND	ND
Fenitrothion	0.030	0.100	ND	ND	ND	ND
Fensulfothion	0.030	0.100	ND	ND	ND	ND
Fenthion	0.030	0.100	ND	ND	ND	ND
Fonofos	0.020	0.100	ND	ND	ND	ND
Azinphos methyl	0.030	0.100	ND	ND	ND	ND
Leptophos	0.030	0.100	ND	ND	ND	ND
Malathion	0.030	0.100	ND	ND	ND	ND
Methidathion	0.030	0.100	ND	ND	ND	ND
Parathion, Methyl	0.010	0.100	ND	ND	ND	ND
Molinate	0.020	0.100	ND	ND	ND	ND
Parathion, Ethyl	0.030	0.100	ND	ND	ND	ND
Phorate	0.050	0.200	ND	ND	ND	ND
Mevinphos	0.030	0.100	ND	ND	ND	ND
Phosmet	0.050	0.200	ND	ND	ND	ND
Phosphamidon	0.030	0.100	ND	ND	ND	ND
Ethoprop	0.030	0.100	ND	ND	ND	ND
Sulfotep	0.030	0.100	ND	ND	ND	ND
Bolstar	0.030	0.100	ND	ND	ND	ND
Terbufos	0.030	0.100	ND	ND	ND	ND
Tetrachlorvinphos	0.030	0.100	ND	ND	ND	ND
Thiobencarb	0.020	0.100	ND	ND	ND	ND
Thionazin	0.040	0.100	ND	ND	ND	ND
Tokuthion	0.030	0.100	ND	ND	ND	ND
Merphos	0.030	0.100	ND	ND	ND	ND
Trichlorfon	0.030	0.100	ND	ND	ND	ND
Trichloronate	0.030	0.100	ND	ND	ND	ND

Table C-17. Concentrations of Pyrethroid Pesticides in Ambersorb and HLB Eluates (μg/L).

Pyrethroids	MDL	RL	Switzer AMB 27	Consolidated AMB	Consolidated HLB	Newport AMB 27	Newport AMB 29	SD Creek AMB 26	SD Creek AMB 27	SD Creek AMB 28	SD Creek AMB 29	Indiana AMB	Indiana HLB	Alisal AMB	Alisal HLB
Bifenthrin	0.0005	0.005		ND	ND	ND	ND	ND	ND	ND	ND	ND	ND	ND	ND
Cyfluthrin-1	0.002	0.022		ND	ND	ND	ND	ND	ND	ND	ND	ND	ND	ND	ND
Cyfluthrin-2	0.002	0.022		ND	ND	ND	ND	ND	ND	ND	ND	ND	ND	ND	ND
Cyfluthrin-3	0.002	0.022		ND	ND	ND	ND	ND	ND	ND	ND	ND	ND	ND	ND
Cyfluthrin-4	0.002	0.022		ND	ND	ND	ND	ND	ND	ND	ND	ND	ND	ND	ND
Cypermethrin-1	0.002	0.022		ND	ND	ND	ND	ND	ND	ND	ND	ND	ND	ND	ND
Cypermethrin-2	0.002	0.022		ND	ND	ND	ND	ND	ND	ND	ND	ND	ND	ND	ND
Cypermethrin-3	0.002	0.022		ND	ND	ND	ND	ND	ND	ND	ND	ND	ND	ND	ND
Cypermethrin-4	0.002	0.022		ND	ND	ND	ND	ND	ND	ND	ND	ND	ND	ND	ND
Esfenvalerate/Fenvalerate-1	0.001	0.010		ND	ND	ND	ND	ND	ND	ND	ND	ND	ND	ND	ND
Esfenvalerate/Fenvalerate-2	0.001	0.010		ND	ND	ND	ND	ND	ND	ND	ND	ND	ND	ND	ND
Lambda-cyhalothrin-1	0.0005	0.005		ND	ND	ND	ND	ND	ND	ND	ND	ND	ND	ND	ND
Lambda-cyhalothrin-2	0.0005	0.005		ND	ND	ND	ND	ND	ND	ND	ND	ND	ND	ND	ND
Permethrin-1	0.003	0.025		ND	ND	ND	ND	ND	ND	ND	ND	ND	ND	ND	ND
Permethrin-2	0.003	0.025		ND	ND	ND	ND	ND	ND	ND	ND	ND	ND	ND	ND
Deltamethrin	0.002	0.025		ND	ND	ND	ND	ND	ND	ND	ND	ND	ND	ND	ND

Table C-18. Concentrations of Organochlorine Pesticides in Ambersorb and HLB Eluates (μg/L).

Organochlorine Pesticides	MDL	RL	Switzer AMB 27	Consolidated AMB	Consolidated HLB	Newport AMB 27	Newport AMB 29	SD Creek AMB 26	SD Creek AMB 27	SD Creek AMB 28	SD Creek AMB 29	Indiana AMB	Indiana HLB	Alisal AMB	Alisal HLB
Aldrin	0.001	0.010	ND	ND	ND	ND	ND	ND	ND	ND	ND	ND	ND	ND	ND
Chlordane, cis-	0.001	0.010	ND	ND	ND	ND	ND	ND	ND	ND	ND	ND	ND	ND	ND
Chlordane, trans-	0.001	0.010	**0.085**	ND	ND	ND	ND	ND	ND	ND	ND	ND	ND	ND	ND
Chlordene, alpha-	0.001	0.010	ND	ND	ND	ND	ND	ND	ND	ND	ND	ND	ND	ND	ND
Chlordene, gamma-	0.001	0.010	ND	ND	ND	ND	ND	ND	ND	ND	ND	ND	ND	ND	ND
Dacthal	0.001	0.010	ND	ND	ND	ND	ND	ND	ND	ND	ND	ND	ND	ND	ND
DDD(o,p')	0.001	0.010	ND	ND	ND	ND	ND	ND	ND	ND	ND	ND	ND	**0.050**	ND
DDD(p,p')	0.001	0.010	ND	ND	ND	ND	ND	ND	ND	ND	ND	ND	ND	ND	**0.210**
DDE(o,p')	0.001	0.010	ND	ND	ND	ND	ND	ND	ND	ND	ND	ND	ND	ND	ND
DDE(p,p')	0.001	0.010	ND	ND	ND	ND	ND	ND	ND	ND	ND	ND	ND	**0.038**	ND
DDMU(p,p')	0.001	0.010	ND	ND	ND	ND	ND	ND	ND	ND	ND	ND	ND	ND	ND
DDT(o,p')	0.001	0.010	ND	ND	ND	ND	ND	ND	ND	ND	ND	ND	ND	ND	ND
DDT(p,p')	0.002	0.025	ND	ND	ND	ND	ND	ND	ND	ND	ND	ND	ND	ND	ND
Dieldrin	0.001	0.010	ND	ND	ND	ND	ND	ND	ND	ND	ND	ND	ND	ND	ND
Endosulfan I	0.001	0.010	ND	ND	ND	ND	ND	ND	ND	ND	ND	ND	ND	ND	ND
Endosulfan II	0.001	0.010	ND	ND	ND	ND	ND	ND	ND	ND	ND	ND	ND	ND	ND
Endosulfan sulfate	0.001	0.010	ND	ND	ND	ND	ND	ND	ND	ND	ND	ND	ND	ND	ND
Endrin	0.001	0.010	ND	ND	ND	ND	ND	ND	ND	ND	ND	ND	ND	ND	ND
Endrin Aldehyde	0.002	0.025	ND	ND	ND	ND	ND	ND	ND	ND	ND	ND	ND	ND	ND
Endrin Ketone	0.002	0.025	ND	ND	ND	ND	ND	ND	ND	ND	ND	ND	ND	ND	ND
HCH, alpha	0.001	0.010	ND	ND	ND	ND	ND	ND	ND	ND	ND	ND	ND	ND	ND
HCH, beta	0.001	0.010	ND	ND	ND	ND	ND	ND	ND	ND	ND	ND	ND	ND	ND
HCH, gamma	0.001	0.010	ND	ND	ND	ND	ND	ND	ND	ND	ND	ND	ND	ND	ND
HCH, delta	0.001	0.010	ND	ND	ND	NA	NA	NA	NA	NA	NA	NA	NA	NA	NA
Heptachlor	0.001	0.010	ND	ND	ND	ND	ND	ND	ND	ND	ND	ND	ND	ND	ND
Heptachlor epoxide	0.001	0.010	ND	ND	ND	ND	ND	ND	ND	ND	ND	ND	ND	ND	ND
Hexachlorobenzene	0.0005	0.005	ND	ND	ND	ND	ND	ND	ND	ND	ND	ND	ND	ND	ND
Methoxychlor	0.001	0.010	ND	ND	ND	ND	ND	ND	ND	ND	ND	ND	ND	ND	ND
Mirex	0.001	0.010	ND	ND	ND	ND	ND	ND	ND	ND	ND	ND	ND	ND	ND
Nonachlor, cis-	0.001	0.010	ND	ND	ND	ND	ND	ND	ND	ND	ND	ND	ND	ND	ND
Nonachlor, trans-	0.001	0.010	ND	ND	ND	ND	ND	ND	ND	ND	ND	ND	ND	ND	ND
Oxadiazon	0.001	0.010	ND	ND	ND	ND	ND	ND	ND	ND	ND	ND	ND	ND	ND
Oxychlordane	0.001	0.010	ND	ND	ND	ND	ND	ND	ND	ND	ND	ND	ND	ND	ND
Tedion	0.001	0.010	ND	ND	ND	ND	ND	ND	ND	ND	ND	ND	ND	ND	ND

Table C-19. Concentrations of Polycyclic Aromatic Hydrocarbons in Ambersorb and HLB Eluates (µg/L).

Polycyclic Aromatic Hydrocarbons	MDL	RL	Switzer AMB 27	Consolidated AMB	Consolidated HLB	Newport AMB 29	Newport AMB 28	SD Creek AMB 26	SD Creek AMB 27	SD Creek AMB 28	SD Creek AMB 29	Indiana AMB	Indiana HLB	Alisal AMB	Alisal HLB
Naphthalene	0.0025	0.005	<RL	0.00383	2.11	NA	NA	NA	NA	NA	NA	0.0089	0.0004	NA	NA
Methylnaphthalene, 2-	0.0025	0.005	<RL	<RL	<RL	NA	NA	NA	NA	NA	NA	0.0230	<RL	NA	NA
Methylnaphthalene, 1-	0.0025	0.005	<RL	<RL	8.16	NA	NA	NA	NA	NA	NA	0.0167	0.0003	NA	NA
Dimethylnaphthalene, 2,6-	0.0025	0.005	<RL	0.00409	9.32	NA	NA	NA	NA	NA	NA	0.0337	0.0004	NA	NA
Trimethylnaphthalene, 2,3,5-	0.0025	0.005	<RL	0.00274	4.89	NA	NA	NA	NA	NA	NA	0.0258	<RL	NA	NA
Naphthalenes, C1 -	0.0025	0.005	<RL	0.00255	9.25	NA	NA	NA	NA	NA	NA	0.0413	0.0007	NA	NA
Naphthalenes, C2 -	0.0025	0.005	<RL	0.04136	51.6	NA	NA	NA	NA	NA	NA	0.1456	0.0031	NA	NA
Naphthalenes, C3 -	0.0025	0.005	<RL	0.09518	110	NA	NA	NA	NA	NA	NA	0.2627	0.0043	NA	NA
Naphthalenes, C4 -	0.0025	0.005	<RL	0.06643	42.7	NA	NA	NA	NA	NA	NA	0.1455	0.0013	NA	NA
Biphenyl	0.0025	0.005	<RL	<RL	0.735	NA	NA	NA	NA	NA	NA	<RL	<RL	NA	NA
Acenaphthylene	0.0025	0.005	<RL	<RL	<RL	NA	NA	NA	NA	NA	NA	0.0027	<RL	NA	NA
Acenaphthene	0.0025	0.005	<RL	<RL	<RL	NA	NA	NA	NA	NA	NA	0.0181	0.0004	NA	NA
Fluorene	0.0025	0.005	<RL	<RL	1.90	NA	NA	NA	NA	NA	NA	0.0114	<RL	NA	NA
Methylfluorene, 1-	0.0025	0.005	<RL	0.00296	7.00	NA	NA	NA	NA	NA	NA	0.0060	0.0013	NA	NA
Fluorenes, C1 -	0.0025	0.005	<RL	0.00849	13.4	NA	NA	NA	NA	NA	NA	0.0285	0.0014	NA	NA
Fluorenes, C2 -	0.0025	0.005	<RL	0.00792	12.3	NA	NA	NA	NA	NA	NA	0.0192	0.0003	NA	NA
Fluorenes, C3 -	0.0025	0.005	<RL	<RL	5.60	NA	NA	NA	NA	NA	NA	<RL	<RL	NA	NA
Dibenzothiophene	0.0025	0.005	<RL	<RL	<RL	NA	NA	NA	NA	NA	NA	0.0028	<RL	NA	NA
Methyldibenzothiophene, 4-	0.0025	0.005	<RL	<RL	1.01	NA	NA	NA	NA	NA	NA	0.0032	<RL	NA	NA
Dibenzothiophenes, C1 -	0.0025	0.005	<RL	<RL	2.19	NA	NA	NA	NA	NA	NA	0.0088	<RL	NA	NA
Dibenzothiophenes, C2 -	0.0025	0.005	<RL	0.00313	2.80	NA	NA	NA	NA	NA	NA	0.0208	0.0005	NA	NA
Dibenzothiophenes, C3 -	0.0025	0.005	<RL	<RL	1.074	NA	NA	NA	NA	NA	NA	0.0179	<RL	NA	NA
Phenanthrene	0.0025	0.005	<RL	<RL	4.90	NA	NA	NA	NA	NA	NA	0.0115	<RL	NA	NA
Methylphenanthrene, 1-	0.0025	0.005	<RL	<RL	2.71	NA	NA	NA	NA	NA	NA	0.0038	<RL	NA	NA
Dimethylphenanthrene, 3,6-	0.0025	0.005	<RL	<RL	0.945	NA	NA	NA	NA	NA	NA	0.0039	<RL	NA	NA
Phenanthrene/Anthracene, C1	0.0025	0.005	<RL	0.00839	15.3	NA	NA	NA	NA	NA	NA	0.0268	0.0003	NA	NA
Phenanthrene/Anthracene, C2	0.0025	0.005	<RL	0.01627	7.98	NA	NA	NA	NA	NA	NA	0.0477	<RL	NA	NA
Phenanthrene/Anthracene, C3	0.0025	0.005	<RL	0.01744	3.95	NA	NA	NA	NA	NA	NA	0.0375	<RL	NA	NA
Phenanthrene/Anthracene, C4	0.0025	0.005	<RL	0.00548	<RL	NA	NA	NA	NA	NA	NA	0.0072	<RL	NA	NA
Anthracene	0.0025	0.005	<RL	<RL	0.922	NA	NA	NA	NA	NA	NA	0.0026	<RL	NA	NA
Fluoranthene	0.0025	0.005	<RL	<RL	<RL	NA	NA	NA	NA	NA	NA	0.0499	<RL	NA	NA
Methylfluoranthene, 2-	0.0025	0.005	<RL	<RL	<RL	NA	NA	NA	NA	NA	NA	0.0047	<RL	NA	NA

Polycyclic Aromatic Hydrocarbons	MDL	RL	Switzer AMB 27	Consolidated AMB	Consolidated HLB	Newport AMB 29	Newport AMB 28	SD Creek AMB 26	SD Creek AMB 27	SD Creek AMB 28	SD Creek AMB 29	Indiana AMB	Indiana HLB	Alisal AMB	Alisal HLB
Fluoranthene/Pyrenes, C1 -	0.0025	0.005	<RL	0.00429	<RL	NA	NA	NA	NA	NA	NA	0.0379	<RL	NA	NA
Pyrene	0.0025	0.005	<RL	<RL	<RL	NA	NA	NA	NA	NA	NA	0.0570	<RL	NA	NA
Benz[a]anthracene	0.0025	0.005	<RL	<RL	<RL	NA	NA	NA	NA	NA	NA	0.0070	<RL	NA	NA
Chrysene	0.0025	0.005	<RL	<RL	<RL	NA	NA	NA	NA	NA	NA	0.0073	<RL	NA	NA
Chrysenes, C1 -	0.0025	0.005	<RL	<RL	<RL	NA	NA	NA	NA	NA	NA	0.0049	<RL	NA	NA
Chrysenes, C2 -	0.0025	0.005	<RL	<RL	<RL	NA	NA	NA	NA	NA	NA	<RL	<RL	NA	NA
Chrysenes, C3 -	0.0025	0.005	<RL	<RL	<RL	NA	NA	NA	NA	NA	NA	<RL	<RL	NA	NA
Benzo(b)fluoranthene	0.0025	0.005	<RL	<RL	<RL	NA	NA	NA	NA	NA	NA	0.0291	<RL	NA	NA
Benzo(k)fluoranthene	0.0025	0.005	<RL	<RL	<RL	NA	NA	NA	NA	NA	NA	0.0082	<RL	NA	NA
Benzo(e)pyrene	0.0025	0.005	<RL	0.00545	0.578	NA	NA	NA	NA	NA	NA	0.0352	<RL	NA	NA
Benzo(a)pyrene	0.0025	0.005	<RL	0.00383	<RL	NA	NA	NA	NA	NA	NA	0.0161	<RL	NA	NA
Perylene	0.0025	0.005	<RL	<RL	<RL	NA	NA	NA	NA	NA	NA	0.0059	<RL	NA	NA
Indeno(1,2,3-c,d)pyrene	0.0025	0.005	<RL	<RL	<RL	NA	NA	NA	NA	NA	NA	<RL	<RL	NA	NA

Table C-20. Concentrations of Polychlorinated Biphenyls in Ambersorb and HLB Eluates (µg/L).

Polychlorinated Biphenyls	MDL	RL	Switzer AMB 27	Consolidated AMB	Consolidated HLB	Newport AMB 27	Newport AMB 29	SD Creek AMB 26	SD Creek AMB 27	SD Creek AMB 28	SD Creek AMB 29	Indiana AMB	Indiana HLB	Alisal AMB	Alisal HLB
PCB 005		0.020	NA	ND	ND	ND	ND	ND	ND	ND	ND	ND	ND	ND	ND
PCB 008		0.020	NA	ND	ND	ND	ND	ND	ND	ND	ND	ND	ND	ND	ND
PCB 015		0.020	NA	ND	ND	ND	ND	ND	ND	ND	ND	ND	ND	ND	ND
PCB 018		0.020	NA	ND	ND	ND	ND	ND	ND	ND	ND	ND	ND	ND	ND
PCB 027		0.020	NA	ND	ND	ND	ND	ND	ND	ND	ND	ND	ND	ND	ND
PCB 028		0.020	NA	ND	ND	ND	ND	ND	ND	ND	ND	ND	ND	ND	ND
PCB 029		0.020	NA	ND	ND	ND	ND	ND	ND	ND	ND	ND	ND	ND	ND
PCB 031		0.020	NA	ND	ND	ND	ND	ND	ND	ND	ND	ND	ND	ND	ND
PCB 033		0.020	NA	ND	ND	ND	ND	ND	ND	ND	ND	ND	ND	ND	ND
PCB 044		0.020	NA	ND	ND	ND	ND	ND	ND	ND	ND	ND	ND	ND	ND
PCB 049		0.020	NA	ND	ND	ND	ND	ND	ND	ND	ND	ND	ND	ND	ND
PCB 052		0.020	NA	ND	ND	ND	ND	ND	ND	ND	ND	ND	ND	ND	ND
PCB 056		0.020	NA	ND	ND	ND	ND	ND	ND	ND	ND	ND	ND	ND	ND
PCB 060		0.020	NA	ND	ND	ND	ND	ND	ND	ND	ND	ND	ND	ND	ND
PCB 066		0.020	NA	ND	ND	ND	ND	ND	ND	ND	ND	ND	ND	ND	ND
PCB 070		0.020	NA	ND	ND	ND	ND	ND	ND	ND	ND	ND	ND	ND	ND
PCB 074		0.020	NA	ND	ND	ND	ND	ND	ND	ND	ND	ND	ND	ND	ND
PCB 087		0.020	NA	ND	ND	ND	ND	ND	ND	ND	ND	ND	ND	ND	ND
PCB 095		0.020	NA	ND	ND	ND	ND	ND	ND	ND	ND	ND	ND	ND	ND
PCB 097		0.020	NA	ND	ND	ND	ND	ND	ND	ND	ND	ND	ND	ND	ND
PCB 099		0.020	NA	ND	ND	ND	ND	ND	ND	ND	ND	ND	ND	ND	ND
PCB 101		0.020	NA	ND	ND	ND	ND	ND	ND	ND	ND	ND	ND	ND	ND
PCB 105		0.020	NA	ND	ND	ND	ND	ND	ND	ND	ND	ND	ND	ND	ND
PCB 110		0.020	NA	ND	ND	ND	ND	ND	ND	ND	ND	ND	ND	ND	ND
PCB 114		0.020	NA	ND	ND	ND	ND	ND	ND	ND	ND	ND	ND	ND	ND
PCB 118		0.020	NA	ND	ND	ND	ND	ND	ND	ND	ND	ND	ND	ND	ND
PCB 128		0.020	NA	ND	ND	ND	ND	ND	ND	ND	ND	ND	ND	ND	ND
PCB 137		0.020	NA	ND	ND	ND	ND	ND	ND	ND	ND	ND	ND	ND	ND
PCB 138		0.020	NA	ND	ND	ND	ND	ND	ND	ND	ND	ND	ND	ND	ND
PCB 141		0.020	NA	ND	ND	ND	ND	ND	ND	ND	ND	ND	ND	ND	ND
PCB 149		0.020	NA	ND	ND	ND	ND	ND	ND	ND	ND	ND	ND	ND	ND
PCB 151		0.020	NA	ND	ND	ND	ND	ND	ND	ND	ND	ND	ND	ND	ND

Polychlorinated Biphenyls	MDL	RL	Switzer AMB 27	Consolidated AMB	Consolidated HLB	Newport AMB 27	Newport AMB 29	SD Creek AMB 26	SD Creek AMB 27	SD Creek AMB 28	SD Creek AMB 29	Indiana AMB	Indiana HLB	Alisal AMB	Alisal HLB
PCB 153		0.020	NA	ND	ND	ND	ND	ND	ND	ND	ND	ND	ND	ND	ND
PCB 156		0.020	NA	ND	ND	ND	ND	ND	ND	ND	ND	ND	ND	ND	ND
PCB 157		0.020	NA	ND	ND	ND	ND	ND	ND	ND	ND	ND	ND	ND	ND
PCB 158		0.020	NA	ND	ND	ND	ND	ND	ND	ND	ND	ND	ND	ND	ND
PCB 170		0.020	NA	ND	ND	ND	ND	ND	ND	ND	ND	ND	ND	ND	ND
PCB 174		0.020	NA	ND	ND	ND	ND	ND	ND	ND	ND	ND	ND	ND	ND
PCB 177		0.020	NA	ND	ND	ND	ND	ND	ND	ND	ND	ND	ND	ND	ND
PCB 180		0.020	NA	ND	ND	ND	ND	ND	ND	ND	ND	ND	ND	ND	ND
PCB 183		0.020	NA	ND	ND	ND	ND	ND	ND	ND	ND	ND	ND	ND	ND
PCB 187		0.020	NA	ND	ND	ND	ND	ND	ND	ND	ND	ND	ND	ND	ND
PCB 189		0.020	NA	ND	ND	ND	ND	ND	ND	ND	ND	ND	ND	ND	ND
PCB 194		0.020	NA	ND	ND	ND	ND	ND	ND	ND	ND	ND	ND	ND	ND
PCB 195		0.020	NA	ND	ND	ND	ND	ND	ND	ND	ND	ND	ND	ND	ND
PCB 200		0.020	NA	ND	ND	ND	ND	ND	ND	ND	ND	ND	ND	ND	ND
PCB 201		0.020	NA	ND	ND	ND	ND	ND	ND	ND	ND	ND	ND	ND	ND
PCB 203		0.020	NA	ND	ND	ND	ND	ND	ND	ND	ND	ND	ND	ND	ND
PCB 206		0.020	NA	ND	ND	ND	ND	ND	ND	ND	ND	ND	ND	ND	ND
PCB 209		0.020	NA	ND	ND	ND	ND	ND	ND	ND	ND	ND	ND	ND	ND

Table C-21. Concentrations of Organophosphate Pesticides in Ambersorb and HLB Eluates (μg/L).

Organophosphate Pesticides	MDL	RL	Switzer AMB 27	Consolidated AMB	Consolidated HLB	Newport AMB 27	Newport AMB 29	SD Creek AMB 26	SD Creek AMB 27	SD Creek AMB 28	SD Creek AMB 29	Indiana AMB	Indiana HLB	Alisal AMB	Alisal HLB
Aspon	0.030	0.100	NA	ND	ND	ND	ND	ND	ND	ND	ND	ND	ND	ND	ND
Azinphos ethyl	0.030	0.100	NA	ND	ND	ND	ND	ND	ND	ND	ND	ND	ND	ND	ND
Carbophenothion	0.030	0.100	NA	ND	ND	ND	ND	ND	ND	ND	ND	ND	ND	ND	ND
Chlorfenvinphos	0.030	0.100	NA	ND	ND	ND	ND	ND	ND	ND	ND	ND	ND	ND	ND
Chlorpyrifos	0.020	0.100	NA	ND	ND	ND	ND	ND	ND	ND	ND	ND	ND	**0.379**	ND
Chlorpyrifos methyl	0.020	0.100	NA	ND	ND	ND	ND	ND	ND	ND	ND	ND	ND	ND	ND
Ciodrin	0.030	0.100	NA	ND	ND	ND	ND	ND	ND	ND	ND	ND	ND	ND	ND
Coumaphos	0.040	0.100	NA	ND	ND	ND	ND	ND	ND	ND	ND	ND	ND	ND	ND
Demeton-s	0.040	0.100	NA	ND	ND	ND	ND	ND	ND	ND	ND	ND	ND	ND	ND
Diazinon	0.005	0.040	NA	ND	ND	ND	ND	ND	ND	ND	ND	ND	ND	**0.440**	ND
Naled	0.030	0.100	NA	ND	ND	ND	ND	ND	ND	ND	ND	ND	ND	ND	ND
Dichlofenthion	0.030	0.100	NA	ND	ND	ND	ND	ND	ND	ND	ND	ND	ND	ND	ND
Dichlorvos	0.030	0.100	NA	ND	ND	ND	ND	ND	ND	ND	ND	ND	ND	ND	ND
Dicrotophos	0.030	0.100	NA	ND	ND	ND	ND	ND	ND	ND	ND	ND	ND	ND	ND
Dimethoate	0.030	0.100	NA	ND	ND	ND	ND	ND	ND	ND	ND	ND	ND	ND	ND
Dioxathion	0.030	0.100	NA	ND	ND	ND	ND	**94**	ND	ND	ND	ND	ND	ND	ND
Disulfoton	0.010	0.100	NA	ND	ND	ND	ND	ND	ND	ND	ND	ND	ND	ND	ND
Ethion	0.020	0.100	NA	ND	ND	ND	ND	ND	ND	ND	ND	ND	ND	ND	ND
Famphur	0.030	0.100	NA	ND	ND	ND	ND	ND	ND	ND	ND	ND	ND	ND	ND
Fenchlorphos	0.030	0.100	NA	ND	ND	ND	ND	ND	ND	ND	ND	ND	ND	ND	ND
Fenitrothion	0.030	0.100	NA	ND	ND	ND	ND	ND	ND	ND	ND	ND	ND	ND	ND
Fensulfothion	0.030	0.100	NA	ND	ND	ND	ND	ND	ND	ND	ND	ND	ND	ND	ND
Fenthion	0.030	0.100	NA	ND	ND	ND	ND	ND	ND	ND	ND	ND	ND	ND	ND
Fonofos	0.020	0.100	NA	ND	ND	ND	ND	ND	ND	ND	ND	ND	ND	ND	ND
Azinphos methyl	0.030	0.100	NA	ND	ND	ND	ND	ND	ND	ND	ND	ND	ND	ND	ND
Leptophos	0.030	0.100	NA	ND	ND	ND	ND	ND	ND	ND	ND	ND	ND	ND	ND
Malathion	0.030	0.100	NA	ND	ND	ND	ND	ND	ND	ND	ND	ND	ND	ND	ND
Methidathion	0.030	0.100	NA	ND	ND	ND	ND	ND	ND	ND	ND	ND	ND	ND	ND
Parathion, Methyl	0.010	0.100	NA	ND	ND	ND	ND	ND	ND	ND	ND	ND	ND	ND	ND
Molinate	0.020	0.100	NA	ND	ND	ND	ND	ND	ND	ND	ND	ND	ND	ND	ND
Parathion, Ethyl	0.030	0.100	NA	ND	ND	ND	ND	ND	ND	ND	ND	ND	ND	ND	ND
Phorate	0.050	0.200	NA	ND	ND	ND	ND	ND	ND	ND	ND	ND	ND	ND	ND

Organophosphate Pesticides	MDL	RL	Switzer AMB 27	Consolidated AMB	Consolidated HLB	Newport AMB 27	Newport AMB 29	SD Creek AMB 26	SD Creek AMB 27	SD Creek AMB 28	SD Creek AMB 29	Indiana AMB	Indiana HLB	Alisal AMB	Alisal HLB
Mevinphos	0.030	0.100	NA	ND	ND	ND	ND	ND	ND	ND	ND	ND	ND	ND	ND
Phosmet	0.050	0.200	NA	ND	ND	ND	ND	ND	ND	ND	ND	ND	ND	ND	ND
Phosphamidon	0.030	0.100	NA	ND	ND	ND	ND	ND	ND	ND	ND	ND	ND	ND	ND
Ethoprop	0.030	0.100	NA	ND	ND	ND	ND	ND	ND	ND	ND	ND	ND	ND	ND
Sulfotep	0.030	0.100	NA	ND	ND	ND	ND	ND	ND	ND	ND	ND	ND	ND	ND
Bolstar	0.030	0.100	NA	ND	ND	ND	ND	ND	ND	ND	ND	ND	ND	ND	ND
Terbufos	0.030	0.100	NA	ND	ND	ND	ND	ND	ND	ND	ND	ND	ND	ND	ND
Tetrachlorvinphos	0.030	0.100	NA	ND	ND	ND	ND	ND	ND	ND	ND	ND	ND	ND	ND
Thiobencarb	0.020	0.100	NA	ND	ND	ND	ND	ND	ND	ND	ND	ND	ND	ND	ND
Thionazin	0.040	0.100	NA	ND	ND	ND	ND	ND	ND	ND	ND	ND	ND	ND	ND
Tokuthion	0.030	0.100	NA	ND	ND	ND	ND	ND	ND	ND	ND	ND	ND	ND	ND
Merphos	0.030	0.100	NA	ND	ND	ND	ND	ND	ND	ND	ND	ND	ND	ND	ND
Trichlorfon	0.030	0.100	NA	ND	ND	ND	ND	ND	ND	ND	ND	ND	ND	ND	ND
Trichloronate	0.030	0.100	NA	ND	ND	ND	ND	ND	ND	ND	ND	ND	ND	ND	ND

Table C-22. Concentrations of Pyrethroid Pesticides in Amphipod Tissues Exposed to Switzer Creek TIE Treatments. Sample Reporting Limit Based on Sample Size.

Pyrethroids (ng/g fresh wt)	Estimated MDL	Method Reporting Limit	Sample Reporting Limit *	SZ SED	SZ AMB	SZ SIR	Control Sediment	Baseline Tissue
Bifenthrin	0.500	1.00	5.00	ND	ND	ND	ND	ND
Cyfluthrin-1	1.50	3.00	15.0	ND	ND	ND	ND	ND
Cyfluthrin-2	1.50	3.00	15.0	ND	ND	ND	ND	ND
Cyfluthrin-3	1.50	3.00	15.0	ND	ND	ND	ND	ND
Cyfluthrin-4	1.50	3.00	15.0	ND	ND	ND	ND	ND
Cypermethrin-1	1.50	3.00	15.0	ND	ND	ND	ND	ND
Cypermethrin-2	1.50	3.00	15.0	ND	ND	ND	ND	ND
Cypermethrin-3	1.50	3.00	15.0	ND	ND	ND	ND	ND
Cypermethrin-4	1.50	3.00	15.0	ND	ND	ND	ND	ND
Deltamethrin	1.00	2.00	10.0	ND	ND	ND	ND	ND
Esfenvalerate/Fenvalerate-1	1.00	2.00	10.0	ND	ND	ND	ND	ND
Esfenvalerate/Fenvalerate-2	1.00	2.00	10.0	ND	ND	ND	ND	ND
Permethrin-1	2.00	4.00	20.0	ND	ND	ND	ND	ND
Permethrin-2	2.00	4.00	20.0	ND	ND	ND	ND	ND
Warrior-1	0.500	1.00	5.00	ND	ND	ND	ND	ND
Warrior-2	0.500	1.00	5.00	ND	ND	ND	ND	ND

Table C-23. Concentrations of Metals in Amphipod Tissues Exposed to Switzer Creek TIE Treatments. Sample Reporting Limit Based on Sample Size.

Pyrethroids (ng/g fresh wt)	Estimated MDL	Method Reporting Limit	SZ SED	SZ AMB	SZ SIR	Control Sediment
Ag – Silver	0.003	0.010	0.038	0.026	0.023	0.029
Al – Aluminum	1.00	3.00	220	46.2	68.7	47.6
As – Arsenic	0.01	0.03	1.17	0.90	0.93	1.10
Cd – Cadmium	0.001	0.003	0.021	0.019	0.021	0.026
Cr – Chromium	0.03	0.10	0.59	0.21	0.24	0.23
Cu – Copper	0.02	0.05	12.1	11.4	10.8	11.9
Mn – Manganese	0.01	0.03	2.59	1.36	1.42	2.64
Ni – Nickel	0.01	0.03	0.40	0.19	0.13	0.19
Pb – Lead	0.003	0.010	0.59	0.23	0.90	0.09
Se – Selenium	0.02	0.06	0.14	0.15	0.14	0.14
Zn – Zinc	0.20	0.60	9.00	7.37	8.21	7.40

Table C-24. Concentrations of Organochlorine Pesticides in Amphipod Tissues Exposed to Switzer Creek TIE Treatments. Sample Reporting Limit Based on Sample Size.

Organochlorine Pesticides (ng/g fresh wt)	Estimated MDL	Method Reporting Limit	Sample Reporting Limit *	SZ SED	SZ AMB	SZ SIR	Control Sediment	Baseline Tissue
Aldrin	0.500	1.00	5.00	ND	ND	ND	ND	ND
Chlordane, cis-	0.500	1.00	5.00	ND	ND	ND	ND	ND
Chlordane, trans-	0.500	1.00	5.00	ND	ND	ND	ND	ND
Chlordene, alpha-	0.500	1.00	5.00	ND	ND	ND	ND	ND
Chlordene, gamma-	0.500	1.00	5.00	ND	ND	ND	ND	ND
Dacthal	0.500	1.00	5.00	ND	ND	ND	ND	ND
DDD(o,p')	0.500	1.00	5.00	ND	ND	ND	ND	ND
DDD(p,p')	0.500	1.00	5.00	ND	ND	ND	ND	ND
DDE(o,p')	0.500	1.00	5.00	ND	ND	**5.14**	ND	ND
DDE(p,p')	0.500	1.00	5.00	ND	ND	ND	ND	ND
DDMU(p,p')	0.500	1.00	5.00	ND	ND	ND	ND	ND
DDT(o,p')	0.500	1.00	5.00	ND	ND	ND	ND	ND
DDT(p,p')	1.00	2.00	10.0	ND	ND	ND	ND	ND
Dieldrin	0.500	1.00	5.00	ND	ND	ND	ND	ND
Endosulfan I	0.500	1.00	5.00	ND	ND	ND	ND	ND
Endosulfan II	0.500	1.00	5.00	ND	ND	ND	ND	ND
Endosulfan sulfate	0.500	1.00	5.00	ND	ND	ND	ND	ND
Endrin	0.500	1.00	5.00	ND	ND	ND	ND	ND
Endrin Aldehyde	1.00	2.00	10.0	ND	ND	ND	ND	ND
Endrin Ketone	1.00	2.00	10.0	ND	ND	ND	ND	ND
HCH, alpha	0.500	1.00	5.00	ND	ND	ND	ND	ND
HCH, beta	0.500	1.00	5.00	ND	ND	ND	ND	ND
HCH, gamma	0.500	1.00	5.00	ND	ND	ND	ND	ND
Heptachlor	0.500	1.00	5.00	ND	ND	ND	ND	ND
Heptachlor epoxide	0.500	1.00	5.00	ND	ND	ND	ND	ND
Hexachlorobenzene	0.250	0.500	2.50	ND	ND	ND	ND	ND
Methoxychlor	0.500	1.00	5.00	ND	ND	ND	ND	ND
Mirex	0.500	1.00	5.00	ND	ND	ND	ND	ND
Nonachlor, cis-	0.500	1.00	5.00	ND	ND	ND	ND	ND
Nonachlor, trans-	0.500	1.00	5.00	ND	ND	ND	ND	ND
Oxadiazon	0.500	1.00	5.00	**28.7**	ND	ND	ND	ND
Oxychlordane	0.500	1.00	5.00	ND	ND	ND	ND	ND
Tedion	0.500	1.00	5.00	ND	ND	ND	ND	ND

Table C-25. Concentrations of Polychlorinated Biphenyls in Amphipod Tissues Exposed to Switzer Creek TIE Treatments. Sample Reporting Limit Based on Sample Size.

Polychlorinated biphenyls (ng/g fresh wt)	Estimated MDL	Method Reporting Limit	Sample Reporting Limit *	SZ SED	SZ AMB	SZ SIR	Control Sediment	Baseline Tissue
PCB 008	0.100	0.200	1.00	ND	ND	ND	ND	ND
PCB 018	0.100	0.200	1.00	ND	ND	ND	ND	ND
PCB 027	0.100	0.200	1.00	ND	ND	ND	ND	ND
PCB 028	0.100	0.200	1.00	ND	ND	ND	ND	ND
PCB 029	0.100	0.200	1.00	ND	ND	ND	ND	ND
PCB 031	0.100	0.200	1.00	ND	ND	ND	ND	ND
PCB 033	0.100	0.200	1.00	ND	ND	ND	ND	ND
PCB 044	0.100	0.200	1.00	ND	ND	ND	ND	ND
PCB 049	0.100	0.200	1.00	ND	ND	ND	ND	ND
PCB 052	0.100	0.200	1.00	ND	ND	ND	ND	ND
PCB 056	0.100	0.200	1.00	ND	ND	ND	ND	ND
PCB 060	0.100	0.200	1.00	ND	ND	ND	ND	ND
PCB 066	0.100	0.200	1.00	ND	ND	ND	ND	ND
PCB 070	0.100	0.200	1.00	ND	ND	ND	ND	ND
PCB 074	0.100	0.200	1.00	ND	ND	ND	ND	ND
PCB 087	0.100	0.200	1.00	ND	ND	ND	ND	ND
PCB 095	0.100	0.200	1.00	ND	ND	ND	ND	ND
PCB 097	0.100	0.200	1.00	ND	ND	ND	ND	ND
PCB 099	0.100	0.200	1.00	ND	ND	ND	ND	ND
PCB 101	0.100	0.200	1.00	ND	ND	ND	ND	ND
PCB 105	0.100	0.200	1.00	ND	ND	ND	ND	ND
PCB 110	0.100	0.200	1.00	ND	ND	ND	ND	ND
PCB 114	0.100	0.200	1.00	ND	ND	ND	ND	ND
PCB 118	0.100	0.200	1.00	ND	ND	ND	ND	ND
PCB 128	0.100	0.200	1.00	ND	ND	ND	ND	ND
PCB 137	0.100	0.200	1.00	ND	ND	ND	ND	ND
PCB 138	0.100	0.200	1.00	ND	ND	ND	ND	ND
PCB 141	0.100	0.200	1.00	ND	ND	ND	ND	ND
PCB 149	0.100	0.200	1.00	ND	ND	ND	ND	ND
PCB 151	0.100	0.200	1.00	ND	ND	ND	ND	ND
PCB 153	0.100	0.200	1.00	ND	ND	ND	ND	ND
PCB 156	0.100	0.200	1.00	ND	ND	ND	ND	ND
PCB 157	0.100	0.200	1.00	ND	ND	ND	ND	ND
PCB 158	0.100	0.200	1.00	ND	ND	ND	ND	ND
PCB 170	0.100	0.200	1.00	ND	ND	ND	ND	ND

Polychlorinated biphenyls (ng/g fresh wt)	Estimated MDL	Method Reporting Limit	Sample Reporting Limit *	SZ SED	SZ AMB	SZ SIR	Control Sediment	Baseline Tissue
PCB 174	0.100	0.200	1.00	ND	ND	ND	ND	ND
PCB 177	0.100	0.200	1.00	ND	ND	ND	ND	ND
PCB 180	0.100	0.200	1.00	ND	ND	ND	ND	ND
PCB 183	0.100	0.200	1.00	ND	ND	ND	ND	ND
PCB 187	0.100	0.200	1.00	ND	ND	ND	ND	ND
PCB 189	0.100	0.200	1.00	ND	ND	ND	ND	ND
PCB 194	0.100	0.200	1.00	ND	ND	ND	ND	ND
PCB 195	0.100	0.200	1.00	ND	ND	ND	ND	ND
PCB 200	0.100	0.200	1.00	ND	ND	ND	ND	ND
PCB 201	0.100	0.200	1.00	ND	ND	ND	ND	ND
PCB 203	0.100	0.200	1.00	ND	ND	ND	ND	ND
PCB 206	0.100	0.200	1.00	ND	ND	ND	ND	ND
PCB 209	0.100	0.200	1.00	ND	ND	ND	ND	ND

Glossary of Terms

Ambersorb 563 – Sediment amendment used to reduce the bioavailability of organic contaminants

Cation Column – Solid-phase extraction column used to remove divalent cationic contaminants from interstitial water

Combination Treatment – Two or more individual TIE treatments that are generally conducted simultaneously. Example: Ambersorb and zeolite addition in a solid-phase TIE.

Eluate – A sample that has been reconstituted from the solvent or acid fraction that has been passed through a solid-phase extraction column

Elute – To pass solvent or acid through a solid-phase extraction column

Formulated Sediment – Prepared sediment used for solid-phase controls, treatment blanks, and dilution blanks.

HLB Column – Solid-phase extraction column used to remove non-polar organic contaminants from interstitial water

Interstitial Water – Water contained in the interstices of the sediment

Powdered Coconut Charcoal (PCC) – Sediment amendment used to reduce the bioavailability of organic contaminants

Rinsate – Interstitial water sample that has been passed through a solid-phase extraction column

Sequential Treatment – Two or more individual TIE treatments that are conducted sequentially. Example: cation-HLB solid-phase extraction. The interstitial water is passed through a cation column then passed through a HLB column.

SIR-300 – Sediment amendment used to reduce the bioavailability of metal contaminants

Solid Phase – Refers to bulk sediment

Toxic Unit – either calculated from an interstitial water dilution series (1 TU = 100/LC50), or used to designate 1 LC50 concentration of a chemical

Zeolite – Sediment amendment used to reduce the concentration of ammonia

REFERENCES

Amweg, E.L., Weston, D.P., Ureda, N.M. 2005. Use and toxicity of pyrethroid pesticides in the Central Valley, CA, U.S. *Environ Toxicol Chem* 24: 966-972.

Anderson, B.S., Hunt, J.W., Phillips, B.M., Fairey, R., Oakden, J., Puckett, H.M., Stephenson, M., Tjeerdema, R.S., Long, E.R., Wilson, C.J., Lyons, M. 2001. Sediment quality in Los Angeles Harbor: A triad assessment. *Environ Toxicol Chem* 20: 359-370.

Anderson, B.S., Nicely, P.A., Phillips, B.M., Hunt, J.W. 2004. Sediment quality assessment study at the B Street/Broadway Piers, Downtown Anchorage, and Switzer Creek, San Diego Bay, California. Technical report, Regional Water Quality Control Board, San Diego, CA. 142 pp with Appendices.

Anderson, B.S., Phillips, B.M., Hunt, J.W., Huntley, S.A., Worcester, K., Richard, N., Tjeerdema, R.S. 2006. Evidence of pesticide impacts in the Santa Maria River watershed (California, U.S.). *Environ Toxicol Chem* 25: 1160-1170.

Ankley, G.T., Schubauer-Berigan, M.K., Dierkes, J.R. 1991. Sediment toxicity identification evaluation: Phase I (characterization), Phase II (identification), and Phase III (confirmation) modifications of effluent procedures. National Effluent Toxicity Assessment Center. Technical Report 08-91.

Ankley, G.T., Schubauer-Berigan, M.K., Monson, P.D. 1995. Influence of pH and hardness on the toxicity of ammonia to the amphipod *Hyalella azteca. Can J Fish Aquat Sci* 52: 2078-2083.

Ankley G.T., Collyard S.A. 1995. Influence of Piperonyl Butoxide on the Toxicity of Organophosphate Insecticides to Three Species of Freshwater Benthic Invertebrates. *Comp Biochem Physiol C* 110: 149-155.

ASTM. 2000. E-1367-99. Standard guide for conducting 10-day static sediment toxicity tests with marine and estuarine amphipods. p. 711-738. In: 2000 Annual Book of ASTM Standards. Vol. 11.05, Philadelphia, PA.

Bagarino, T. 1992. Sulfide as an environmental factor and toxicant: tolerance and adaptations in aquatic organisms. *Aquat Toxicol* 24: 21-62.

Bay, S.M., Anderson, B.S., Carr, R.S. 2003. Relative performance of interstitial water and solid-phase toxicity tests: characteristics, causes, and consequences. In: Carr R.S., Nipper M., Eds., *Pore water toxicity testing: biological chemical and ecological considerations – methods, applications, and recommendations for future areas of research*. SETAC Press, Pensacola, FL. Pp 11-34.

Besser, J.M., Ingersoll, C.G., Leonard, E.N., Mount, D.R. 1998. Effect of zeolite on toxicity of ammonia in freshwater sediments: implications for toxicity identification evaluation procedures. *Environ Toxicol Chem* 17: 2310-2317.

Boese, B.L., Lamberson, J.O., Swartz, R.C., Ozretich, R.J. 1997. Photoinduced toxicity of fluoranthene to seven marine benthic crustaceans. *Arch Environ Contam Toxicol* 32: 389–393.

Borgmann, U., Norwood, W.P., Ralph, K.M. 1990. Chronic toxicity and bioaccumulation of 2,3,2',5'- and 3,4,3',4'-tetrachlorobiphenyl and Aroclor® 1242 in the amphipod *Hyalella azteca*. *Arch Environ Contam Toxicol* 19: 558-564.

Brandt, A., Scharf, M., Pedra, J.H.F., Holmes, G., Dean, A., Kreitman, M., Pittendrigh, B.R. 2002. Differential expression and induction of two *Drosophila* cytochrome P450 genes near the Rst(2)DDT locus. *Insect Molecular Biology* 11: 337-341.

Brooke, L.T. 1993. Acute and chronic toxicity of nonylphenol to ten species of aquatic organisms. Contract No. 68-C1-0034, U.S EPA, Duluth, MN, 36p.

Brown, M.A. 1987. Temperature-dependent pyrethroid resistance in a pyrethroid-selected colony of *Heliothis virescens*. *J Econ Entomol* 80: 330-332.

Burgess, R.M., Charles, J.B., Kuhn, A., Ho, K.T., Patton, L., McGovern, D. 1997. Development of cation-exchange methodology for marine toxicity identification evaluation applications. *Environ Toxicol Chem* 16: 1203-1211.

Burgess, R.M., Cantwell, M.G., Pelletier, M.C., Ho, K.T., Serbst, J.R., Cook, H., Kuhn, A. 2000. Development of a toxicity identification evaluation procedure for characterizing metal toxicity in marine sediments. *Environ Toxicol Chem* 19: 982-991.

Burgess, R.M., Pelletier, M.C., Ho, K.T., Serbst, J.R., Ryba, S.A., Kuhn, A., Perron, M.M., Raczelowski, P., Cantwell, M.G. 2003. Removal of ammonia toxicity in marine sediment TIEs: a comparison of *Ulva lactuca*, zeolite and aeration methods. *Mar Poll Bull* 46: 607-618.

Burgess, R.M., Perron, M.M., Cantwell, M.G., Ho, K.T., Serbst, J.R., Pelletier, M.C. 2004. Use of zeolite for removing ammonia and ammonia-caused toxicity in marine toxicity identification evaluations. *Arch Environ Contam Toxicol* 47: 440-447.

Burgess, R.M., Perron, M.M., Cantwell, M.G., Ho, K.T., Pelletier, M.C., Serbst, J.R., Ryba, S.A. 2007. Marine sediment toxicity identification evaluation methods for the anionic metals arsenic and chromium. *Environ Toxicol Chem* 26:61-67.

Coats, J.R., Symonik, D.M., Bradbury, S.P., Dyer, S.D., Timson, L.K., Atchison, G.J. 1989. Toxicology of synthetic pyrethroids in aquatic organisms: an overview. *Environ Toxicol Chem* 8: 671-679.

Cole, F.A., Boese, B.L., Swartz, R.C., Lamberson, J.O., DeWitt, T.H. 2000. Effects of storage on the toxicity of sediments spiked with fluoranthene to the amphipod, *Rhepoxynius abronius*. *Environ Toxicol Chem* 19: 744-748.

Cornelissen, G., Rigternink, H., van Noort, P.C.M., Govers, H.A.J. 2000. Slowly and very slowly desorbing organic compounds in sediments exhibit Langmuir-type sorption. *Environ Toxicol Chem* 19: 1532-1540.

DeWitt, T.H., Ditsworth, G.R., Swartz, R.C. 1988. Effects of natural sediment features on survival of the phoxocephalid amphipod, *Rhepoxynius abronius*. *Mar Environ Res* 25: 99-124.

DeWitt, T.H., Swartz, R.C., Lamberson, J.O. 1989. Measuring the acute toxicity of estuarine sediments. *Environ Toxicol Chem* 8: 1035-1048.

Di Toro, D.M., McGrath, J.A., Hansen, D.J. 2000. Technical basis for narcotic chemicals and polycyclic aromatic hydrocarbon criteria. I. Water and Tissue. *Environ Toxicol Chem* 19: 1951-1970.

Di Toro, D.M., McGrath, J.A. 2000. Technical basis for narcotic chemicals and polycyclic aromatic hydrocarbon criteria. II. Mixtures and Sediments. *Environ Toxicol Chem* 19: 1971-1982.

Ditsworth, G.R., Schults, D.W., Jones, J.K.P. 1990. Preparation of benthic substrates for sediment toxicity testing. *Environ Toxicol Chem* 9: 1523-1529.

Doe, K.G., Burton, G.A., Ho, K.T. 2003. Interstitial water toxicity testing: an overview. In: Carr RS, Nipper M, Eds., *Interstitial water toxicity testing: biological, chemical, and ecological considerations – methods, applications, and recommendations for future areas of research.* Society of Environmental Toxicology and Chemistry Press, Pensacola, FL. Pp 125-142.

Fay, A.A., Brownawell, B.J., Elskus, A.A., McElroy, A.E. 2000. Critical body residues in the marine amphipod *Ampelisca abdita*: sediment exposures with nonionic organic contaminants. *Environ Toxicol Chem* 19: 1028-1035.

Goodfellow, W.L., Ausley, L.W., Burton, D.T., Denton, D.L., Dorn, P.B., Grothe, D.R., Heber, M.A., Norberg-King, T.J., Rodgers, J.H. 2000. Major ion toxicity in effluents: a review with permitting recommendations. *Environ Toxicol Chem* 19: 175-182.

Hatch, A.C., Burton, Jr., G.A. 1999. Photo-Induced Toxicity of PAHs to *Hyalella azteca* and *Chironomus tentans*: Effects of Mixtures and Behavior. *Environ Poll* 106: 157-167.

Hecht, S., Boese, B.L. 2002. Sensitivity of an infaunal amphipod, *Eohaustorius estuarius*, to acute waterborne exposures of 4-nonylphenol: evidence of a toxic hangover. *Environ Toxicol Chem* 21: 816-819.

Heinis, L.J., Highland, T.L., Mount, D.R. 2004. Method for Testing the Aquatic Toxicity of Sediment Extracts for Use in Identifying Organic Toxicants in Sediments. *Environ. Sci Technol* 38: 6256-6262

Ho, K.T., Kuhn, A., Pelletier, M.C., Hendricks, T.L., Helmstetter, A. 1999. pH dependent toxicity of five metals to three marine organisms. *Environ Toxicol* 14: 235-240.

Ho, K.T., Kuhn, A., Pelletier, M.C., Serbst, J.R., Ryba, S.A., Cantwell, M.G., Kuhn, A., Raczelowski, P. 2002. An overview of toxicant identification in sediments and dredged materials. *Mar Poll Bull* 44: 286-293.

Ho, K.T., Kuhn, A., Pelletier, M.C., Serbst, J.R., Cook, H., Cantwell, M.G., Ryba, S.A., Perron, M.M., Lebo, J., Huckins, J., Petty, J. 2004. Use of powdered coconut charcoal as a toxicity identification and evaluation manipulation for organic toxicants in marine sediments. *Environ Toxicol Chem* 23: 2124-2131.

Hoke, R.A., Ankley, G.T., Cotter, A.M., Goldenstein, T., Kosian, P.A., Phipps, G.L., VanderMeiden, F.M. 1994. Evaluation of equilibrium partitioning theory for predicting acute toxicity of field-collected sediments contaminated with DDT, DDE, and DDD to the amphipod *Hyalella azteca*. *Environ Toxicol Chem* 13: 157-166.

Hunt, J.W., Anderson, B.S., Phillips, B.M., Tjeerdema, R.S., Taberski, K.M., Wilson, C.J., Puckett, H.M., Stephenson, M., Fairey, R., Oakden, J. 2001. A large-scale categorization of sites in San Francisco Bay, U.S., based on the sediment quality triad, toxicity identification evaluations, and gradient studies. *Environ Toxicol Chem* 20: 1252-1265.

Hunt, J.W., Anderson, B.S., Phillips, B.M., Tjeerdema, R.S., Richard, N., Connor, V., Worcester, K., Angelo, M., Bern, A., Fulfrost, B., Mulvaney, D. 2006. Spatial relationships between water quality and pesticide applications in agricultural watersheds. *Environ Monit Assess* 121: 245-262.

Ingersoll, C.G., MacDonald, D.D., Brumbaugh, W.G., Johnson, B.T., Kemble, N.E., Kunz, J.L., May, T.W., Wang, N., Smith, J.R., Sparks, D.W., Ireland, D.S. 2002. Toxicity assessment of sediments from the Grand Calumet River and Indiana Harbor Canal in Northwestern Indiana, U.S. *Arch Environ Contam Toxicol* 43: 156-167.

Kakko, I., Toimela, T., Tähti, H. 2000. Piperonyl butoxide potentiates the synaptosome ATPase inhibiting effect of pyrethrin. *Chemosphere* 40: 301-305.

Kesraoui-Ouki, S., Cheeseman, C.R., Perry, R. 1994. Natural zeolite utilization in pollution control: a review of applications to metals' effluents. *J Chem Technol Biotechnol* 59: 121-126.

Knezovich, J.P., Steichen, D.J., Jelinski, J.A., Anderson, S.L. 1996. Sulfide tolerance of four marine species used to evaluate sediment and pore water toxicity. *Bull Environ Contam Toxicol* 57: 450-457.

Kosian, P.A., West, C.W., Pasha, M.S., Cox, J.S., Mount, D.R., Huggett, R.J., Ankley, G.T. 1999. Use of nonpolar resin for reduction of fluoranthene bioavailability in sediment. *Environ Toxicol Chem* 18: 201-206.

Landrum, P.F., Steevens, J.A., McElroy, M., Gossiaux, D.C., Lewis, J.S., Robinson, S.D. 2005. Time-dependent toxicity of dichlorodiphenyldichloroethylene to *Hyalella azteca*. *Environ Toxicol Chem* 24: 211-218.

Lasier, P.J., Winger, P.V., Reinert, R.E. 1997. Toxicity of alkalinity to *Hyalella azteca*. *Bull Environ Contam Toxicol* 59: 807-814.

Lasier, P.J., Winger, P.V., Bogenrieder, K.J. 2000. Toxicity of manganese to *Ceriodaphnia dubia* and *Hyalella azteca*. *Arch Environ Contam Toxicol* 38: 298-304.

Lebo, J.A., Huckins, J.N., Petty, J.D., Ho, K.T. 1999. Removal of organic contaminant toxicity from sediments--early work toward development of a toxicity identification evaluation (TIE) method. *Chemosphere* 39: 389-406.

Lebo, J.A., Huckins, J.N., Petty, J.D. 2000. Selective removal of organic contaminants from sediments: a methodology for toxicity identification evaluations (TIEs). *Chemosphere* 40: 811-819

Lebo, J.A., Huckins, J.N., Petty, J.D., Cranor, W.L., Ho, K.T. 2003. Comparisons of coarse and fine versions of two carbons for reducing the bioavailabilities of sediment-bound hydrophobic organic contaminants. *Chemosphere* 50: 1309-1317.

Long, E.R., Macdonald, D.D., Smith, S.L., Calder, F.D. 1995. Incidence of adverse biological effects within ranges of chemical concentrations in marine and estuarine sediments. *Environmental Management* 19: 81-97.

Long, E.R., Macdonald, D.D. 1998. Recommended uses of empirically derived, sediment quality guidelines for marine and estuarine ecosystems. *Human Ecolog Risk Assess* 4: 1019-1039.

Los Angeles Regional Water Quality Control Board (LARWQCB). 2004. Watershed Management Initiative Chapter: Dominquez Channel and Los Angeles/Long Beach Harbors Watershed Management Areas. California Regional Water Quality Control Board, Los Angeles Region. pp 181 w/Appendices.

Macdonald, D.D., Dipinto, L.M., Field, J., Ingersoll, C.G., Long, E.R., Swartz, R.C. 2000a. Development and evaluation of consensus-based sediment effect concentrations for polychlorinated biphenyls. *Environ Toxicol Chem* 19 (5): 1403-1413.

Macdonald, D.D., Ingersoll, C.G., Berger, T.A. 2000b. Development and evaluation of consensus-based sediment quality guidelines for freshwater ecosystems. *Arch Environ Contam Toxicol* 39 (1): 20-31.

Macdonald, D.D., Ingersoll, C.G., Smorong, D.E., Lindskoog, R.A., Sparks, D.W., Smith, J.R., Simon, T.P., Hanacek, M.A. 2002. An assessment of injury to sediments and sediment-dwelling organisms in the Grand Calumet River and Indiana Harbor area of concern, U.S. *Arch Environ Contam Toxicol* 43: 141-155.

Malueg, K.W., Schuytema, G.S., Krawczyk, D.F. 1986. Effects of sample storage on a copper-spiked freshwater sediment. *Environ Toxicol Chem* 5: 245-253.

Mayer, F.L., Ellersick, M.R. 1986. Manual of acute toxicity: Interpretation and data base for 410 chemicals and 66 species of freshwater animals. Resource Publication No. 160, U.S. Fish and Wildlife Service, Washington, D.C.

McCarty, L.S., Mackay, D. 1993. Enhancing ecotoxicological modeling and assessment. *Environ Sci Technol* 27: 1719–1728.

Moore, D.W., Diener, D., Hester, B., Anghera, S., Farrar, D., Bridges, T. 2004. Field validation of chronic sublethal dredged material laboratory bioassays: a 3 year summary. Proceedings, SETAC, National Meeting, Portland, OR, U.S. p 30.

Nebeker, A.V., Schuytema, G.S., Griffis, W.L., Barbitta, J.A., Carey, L.A. 1989. Effect of sediment organic carbon on survival of *Hyalella azteca* exposed to DDT and endrin. *Environ Toxicol Chem* 8: 705-718.

NOAA (National Oceanic and Atmospheric Administration). 1989. Standard analytical procedures of the NOAA Analytical Facility. 2nd ed. NOAA Technical Memorandum NMFS F/NWC-92, 1985-86. National Status and Trends Program, U.S. Department of Commerce, Rockville, MD.

Norberg-King TJ, Ausley LW, Burton DT, Goodfellow WL, Miller JL, Waller WT. 2005. Toxicity reduction and toxicity identification evaluations for effluents, ambient waters, and other aqueous media. Pensacola FL, USA: Society of Environmental Toxicology and Chemistry (SETAC). 469 p.

Page, D.S., Boehm, P.D., Stubblefield, W.A., Parker, K.R., Gilfillan, E.S., Neff, J.M., Maki, A.W. 2002. Hydrocarbon composition and toxicity of sediment following the Exxon Valdez oil spill in Prince William Sound, Alaska, U.S. *Environ Toxicol Chem* 21: 1438-1450.

Phillips, B.M., Anderson, B.S., Hunt, J.W. 1997. Measurement and distribution of interstitial and overlying water ammonia and hydrogen sulfide in sediment toxicity tests. *Marine Environmental Research.* 44: 117-126.

Phillips, B.M., Anderson, B.S., Hunt, J.W., Nicely, P.A., Kosaka, R., de Vlaming, V., Connor, V., Richard, N., Tjeerdema, R.S. 2004. *In situ* water and sediment toxicity in an agricultural watershed. *Environ Toxicol Chem* 23: 435-442.

Phipps, G.L., Mattson, V.R., Ankley, G.T. 1995. The relative sensitivity of three benthic test species to ten chemicals. *Arch Environ Toxicol Chem* 28: 281-286.

Robinson, A.M., Lamberson, J.O., Cole, F.A., Swartz, R.C. 1988. Effects of culture conditions on the sensitivity of the phoxocephalid amphipod, *Rhepoxynius abronius*, to cadmium in sediment. *Environ Toxicol Chem* 7: 953-959

Science Applications International Corporation (SAIC). 2003. Guide for planning and conducting sediment pore water toxicity identification evaluations (TIE) to determine causes of acute toxicity at navy aquatic sites. User's Guide UG-2052-ENV. Naval Facilities Engineering Service Center, Port Hueneme, CA, U.S. Prepared by Science Applications International Corporation, Newport, RI.

Simpson, S.L., Angel, B.M., Jolley, D.F. 2004. Metal equilibration in laboratory-contaminated (spiked) sediment used for the development of whole-sediment toxicity tests. *Chemosphere* 54: 597-609.

Suedel, B.C., Rodgers, J.H., Clifford, P.A. 1993. Bioavailability of fluoranthene in freshwater sediment toxicity tests. *Environ Toxicol Chem* 12: 155-165.

Sullivan, J.J., Goh, K.S. 2000. Evaluation and validation of a commercial ELISA for diazinon in surface waters. *J Agric Food Chem* 48: 4071-4078.

Swartz, R.C. 1999. Consensus sediment quality guidelines for polycyclic aromatic hydrocarbon mixtures. *Environ Toxicol Chem* 18: 780-787.

Swartz, R.C., Schults, D.W., Ozretich, R.J., Lamberson, J.D., Cole, F.A., DeWitt, T.H., Redmond, M.S., Ferraro, S.P. 1995. Sum PAH: a model to predict the toxicity of polynuclear aromatic hydrocarbon mixtures in field-collected sediments. *Environ Toxicol Chem* 14: 1977-1987.

Swartz, R.C., Cole, F.A., Lamberson, J.O., Ferraro, S.P., Schults, D.W., DeBen, W.A., Lee, H., Ozretich, R.J. 1994. Sediment toxicity, contamination and amphipod abundance at a DDT- and dieldrin-contaminated site in San Francisco Bay. *Environ Toxicol Chem* 13: 949-962.

Tay, K.-L., Doe, K., Jackman, P., MacDonald, A. 1998. Assessment and evaluation of the effects of particle size, ammonia, and sulfide on the acute lethality test. Technical report, Environment Canada, Atlantic Region.

U.S. Environmental Protection Agency. 1990. CLP-M: Inductively Coupled Plasma-Mass Spectrometry. EPA method 6020. 01A0007861. Washington, D.C.

U.S. Environmental Protection Agency. 1991. Methods for aquatic toxicity identification evaluations. Phase I Toxicity Characterization Procedures. EPA 600/6-91/003. Office of Research and Development. Washington, D.C.

U.S. Environmental Protection Agency. 1993a. Methods for aquatic toxicity identification evaluations. Phase II Toxicity Identification Procedures for Samples Exhibiting Acute and Chronic Toxicity. EPA 600/R-92/080. Office of Research and Development. Washington, D.C.

U.S. Environmental Protection Agency. 1993b. Methods for aquatic toxicity identification evaluations. Phase III Confirmation Procedures for Samples Exhibiting Acute and Chronic Toxicity. EPA 600/R-92/081. Office of Research and Development. Washington, D.C.

U.S. Environmental Protection Agency. 1993c. Methods for the Determination of Non-conventional Pesticides in Municipal and Industrial Wastewater, Volume I. Revision 1. EPA 821/R-93-010-A. Office Water, Washington, D.C.

U.S. Environmental Protection Agency. 1994a. Determination of trace elements in water and waste by inductively coupled–mass spectroscopy. EPA method 200.8, revision 1. Technical Report. EPA/600/R-94/111. Washington, D.C.

U.S. Environmental Protection Agency. 1994b. Test methods for evaluating solid waste, physical/chemical methods, 3rd Edition. EPA SW-846. Office of Solid Waste and Emergency Response, Washington, D.C.

U.S. Environmental Protection Agency. 1994c. Methods for assessing the toxicity of sediment-associated contaminants with estuarine and marine amphipods. EPA/600/R-94/025. Office of Research and Development, Washington, D.C.

U.S. Environmental Protection Agency. 1995. Method 1640: Determination of Trace Elements in Ambient Waters by On-Line Chelation Pre-concentration and Inductively Coupled Plasma-Mass Spectrometry. U.S. Environmental Protection Agency (USEPA), Office of Water. EPA 821-R-95-033.

U.S. Environmental Protection Agency. 1996. Marine Toxicity Identification Evaluation (TIE), Phase I Guidance Document. Office of Research and Development. EPA/600/R-95/054. Washington, D.C.

U.S. Environmental Protection Agency. 2000. Methods for measuring the toxicity and bioaccumulation of sediment-associated contaminants with freshwater invertebrates. EPA/600/R-99/064. Office of Research and Development, Washington, D.C.

U.S. Environmental Protection Agency. 2003. Procedures for the Derivation of Equilibrium Partitioning Sediment Benchmarks (ESBs) for the Protection of Benthic Organisms: Dieldrin. EPA-600-R-02 010. Office of Research and Development. Washington, D.C.

U.S. Environmental Protection Agency. 2004. Incidence and severity of sediment contamination in surface waters of the United States. National Sediment Quality Survey; Second Edition. EPA-823-R-04-007. Office of Science and Technology, Standards and Health Protection Division, Washington, D.C. 280 pp.

U.S. Environmental Protection Agency. In press. Sediment Toxicity Identification Evaluation (TIE) Phases I, II, and III Guidance Document. Draft EPA # 600 xxx xxx. Office of Research and Development, Atlantic Ecology Division. Narragansett, RI

Waller, W.T., Bailey, H.A., DeVlaming, V., Ho, K.T., Hunt, J.W., Miller, J.L., Pillard, D.A., Rowland, C.D., Venables, B.J. 2005. Ambient water, interstitial water, and sediment. In Norberg-King, T.J., Ausley, L.W., Burton, D.T., Miller, J.L., Waller, W.T., Eds. *Toxicity reduction and toxicity identification evaluation for effluents, ambient waters, and other aqueous media*. Society of Environmental Toxicology and Chemistry Press, Pensacola, FL. Pp 93-114.

Ware, G.W., 1989. The Pesticide Book. Thompson Publications, Fresno, CA.

Waters Corporation. 2001. Waters Oasis Sample Extraction Products: Environmental and Agrochemical Applications Notebook. Waters Corporation, Milford, MA.

West, C.W., Kosian, P.A., Mount, D.R., Makynen, E.A., Pasha, M.S., Sibley, P.K., Ankley, G.T. 2001. Amendment of sediments with a carbonaceous resin reduces bioavailability of polycyclic aromatic hydrocarbons. *Environ Toxicol Chem* 20: 1104-1111.

Weston, D.P. 2006. Temperature dependence of pyrethroid toxicity: TIE applications and environmental consequences. Proceedings, SETAC, National Meeting, Montreal, QC, Canada. p 26.

Wheelock, C.E., Miller, J.L., Miller, M.J., Gee, S.J., Shan, G., Hammock, B.D. 2004. Development of toxicity identification evaluation procedure for pyrethroid detection using esterase activity. *Environ Toxicol Chem* 23: 2699-2708.

Wheelock, C.E., Miller, J.L., Miller, M.J., Phillips, B.M., Gee, S.J., Tjeerdema, R.S., Hammock, B.D. 2005. Influence of container adsorption upon observed pyrethroid toxicity to *Ceriodaphnia dubia* and *Hyalella azteca*. *Aquat Toxicol* 74: 47-52.

Wheelock, C., Miller, J.L., Miller, M.J., Phillips, B.M., Huntley, S.A., Gee, S.J., Tjeerdema, R.S., Hammock, B.D. 2006. Use of carboxylesterase activity to remove pyrethroid-associated toxicity to *Ceriodaphnia dubia* and *Hyalella azteca* in toxicity identification evaluations (TIEs). *Environ Toxicol Chem* 25: 973-984.

Williamson, B., Burgess, R.M. 2003. Sediment and Interstitial water Chemistry. In: Carr, RS, Nipper, M, Eds., *Interstitial water toxicity testing: biological, chemical, and ecological considerations – methods, applications, and recommendations for future areas of research*. Society of Environmental Toxicology and Chemistry Press, Pensacola, FL. Pp 63-94.

Young, D.L., Bay, S.M., Greenstein, D.J., Anderson, B., Phillips, B., Ringwood, A., Keppler, C., Hester, B. 2005. Inter-laboratory variation and applicability of sublethal sediment test methods using *Mercenaria mercenaria*, and *Mytilus galloprovincialis*. Proceedings, SETAC, National Meeting, Baltimore, MD. p 319.

WASTEWATER UTILITY

Alabama
Montgomery Water Works & Sanitary Sewer Board

Alaska
Anchorage Water & Wastewater Utility

Arizona
Glendale, City of, Utilities Department
Mesa, City of
Peoria, City of
Phoenix Water Services Dept.
Pima County Wastewater Management
Safford, City of

Arkansas
Little Rock Wastewater Utility

California
Central Contra Costa Sanitary District
Corona, City of
Crestline Sanitation District
Delta Diablo Sanitation District
Dublin San Ramon Services District
East Bay Dischargers Authority
East Bay Municipal Utility District
Eastern Municipal Water District
El Dorado Irrigation District
Fairfield-Suisun Sewer District
Fresno Department of Public Utilities
Inland Empire Utilities Agency
Irvine Ranch Water District
Las Virgenes Municipal Water District
Livermore, City of
Los Angeles, City of
Los Angeles County, Sanitation Districts of
Napa Sanitation District
Orange County Sanitation District
Palo Alto, City of
Riverside, City of
Sacramento Regional County Sanitation District
San Diego Metropolitan Wastewater Department, City of
San Francisco, City & County of
San Jose, City of
Santa Barbara, City of
Santa Cruz, City of
Santa Rosa, City of
South Bayside System Authority
South Coast Water District
South Orange County Wastewater Authority
Stege Sanitary District
Sunnyvale, City of
Union Sanitary District
West Valley Sanitation District

Colorado
Aurora, City of
Boulder, City of
Greeley, City of
Littleton/Englewood Water Pollution Control Plant
Metro Wastewater Reclamation District, Denver

Connecticut
Greater New Haven WPCA
Stamford, City of

District of Columbia
District of Columbia Water & Sewer Authority

Florida
Broward, County of
Fort Lauderdale, City of
Miami-Dade Water & Sewer Authority
Orange County Utilities Department
Reedy Creek Improvement District
Seminole County Environmental Services
St. Petersburg, City of
Tallahassee, City of
Tampa, City of
Toho Water Authority
West Palm Beach, City of

Georgia
Atlanta Department of Watershed Management
Augusta, City of
Clayton County Water Authority
Cobb County Water System
Columbus Water Works
Fulton County
Gwinnett County Department of Public Utilities
Savannah, City of

Hawaii
Honolulu, City & County of

Idaho
Boise, City of

Illinois
American Bottoms Wastewater Treatment Plant
Greater Peoria Sanitary District
Kankakee River Metropolitan Agency
Metropolitan Water Reclamation District of Greater Chicago
Wheaton Sanitary District

Iowa
Ames, City of
Cedar Rapids Wastewater Facility
Des Moines, City of
Iowa City

Kansas
Johnson County Unified Wastewater Districts
Unified Government of Wyandotte County/ Kansas City, City of

Kentucky
Louisville & Jefferson County Metropolitan Sewer District
Sanitation District No. 1

Louisiana
Sewerage & Water Board of New Orleans

Maine
Bangor, City of
Portland Water District

Maryland
Anne Arundel County Bureau of Utility Operations
Howard County Department of Public Works
Washington Suburban Sanitary Commission

Massachusetts
Boston Water & Sewer Commission
Massachusetts Water Resources Authority (MWRA)
Upper Blackstone Water Pollution Abatement District

Michigan
Ann Arbor, City of
Detroit, City of
Holland Board of Public Works
Saginaw, City of
Wayne County Department of Environment
Wyoming, City of

Minnesota
Rochester, City of
Western Lake Superior Sanitary District

Missouri
Independence, City of
Kansas City Missouri Water Services Department
Little Blue Valley Sewer District
Metropolitan St. Louis Sewer District

Nebraska
Lincoln Wastewater System

Nevada
Henderson, City of
Reno, City of

New Jersey
Bergen County Utilities Authority
Ocean County Utilities Authority
Passaic Valley Sewerage Commissioners

New York
New York City Department of Environmental Protection

North Carolina
Charlotte/Mecklenburg Utilities
Durham, City of
Metropolitan Sewerage District of Buncombe County
Orange Water & Sewer Authority

Ohio
Akron, City of
Butler County Department of Environmental Services
Columbus, City of
Metropolitan Sewer District of Greater Cincinnati
Northeast Ohio Regional Sewer District
Summit, County of

Oklahoma
Oklahoma City Water & Wastewater Utility Department
Tulsa, City of

Oregon
Albany, City of
Clean Water Services
Eugene, City of
Gresham, City of
Portland, City of Bureau of Environmental Services
Water Environment Services

Pennsylvania
Philadelphia, City of
University Area Joint Authority

South Carolina
Charleston Water System
Mount Pleasant Waterworks & Sewer Commission
Spartanburg Water

Tennessee
Cleveland, City of
Knoxville Utilities Board
Murfreesboro Water & Sewer Department
Nashville Metro Water Services

Texas
Austin, City of
Dallas Water Utilities
Denton, City of
El Paso Water Utilities
Fort Worth, City of
Houston, City of
San Antonio Water System
Trinity River Authority

Utah
Salt Lake City Corporation

Virginia
Alexandria Sanitation Authority
Arlington, County of

Fairfax County
Hampton Roads Sanitation District
Hanover, County of
Henrico, County of
Hopewell Regional Wastewater Treatment Facility
Loudoun County Sanitation Authority
Lynchburg Regional WWTP
Prince William County Service Authority
Richmond, City of
Rivanna Water & Sewer Authority

Washington

Everett, City of
King County Department of Natural Resources
Seattle Public Utilities
Sunnyside, Port of
Yakima, City of

Wisconsin

Green Bay Metro Sewerage District
Kenosha Water Utility
Madison Metropolitan Sewerage District
Milwaukee Metropolitan Sewerage District
Racine, City of
Sheboygan Regional Wastewater Treatment
Wausau Water Works

Australia

ACTEW (Ecowise)
South Australian Water Corporation
Sydney Water Corporation
Water Corporation of Western Australia

Canada

Lethbridge, City of
Regina, City of, Saskatchewan
Toronto, City of, Ontario
Winnipeg, City of, Manitoba

New Zealand

Watercare Services Limited

United Kingdom

United Utilities North West (UUNW)

STORMWATER UTILITY

California

Fresno Metropolitan Flood Control District
Los Angeles, City of, Department of Public Works
Monterey, City of
Sacramento, County of
San Francisco, City & County of
Santa Rosa, City of
Sunnyvale, City of

Colorado

Aurora, City of
Boulder, City of

Georgia

Griffin, City of

Iowa

Cedar Rapids Wastewater Facility
Des Moines, City of

Kansas

Overland Park, City of

Kentucky

Louisville & Jefferson County Metropolitan Sewer District

Maine

Portland Water District

North Carolina

Charlotte, City of, Stormwater Services

Pennsylvania

Philadelphia, City of

Tennessee

Chattanooga Stormwater Management

Texas

Harris County Flood Control District, Texas

Washington

Bellevue Utilities Department
Seattle Public Utilities

STATE

Arkansas Department of Environmental Quality
Connecticut Department of Environmental Protection
Kansas Department of Health & Environment
Kentucky Department of Environmental Protection
New England Interstate Water Pollution Control Commission (NEIWPCC)
Ohio River Valley Sanitation Commission
Urban Drainage & Flood Control District, CO

CORPORATE

ADS Environmental Services
Advanced Data Mining International
Alan Plummer & Associates
Alpine Technology Inc.
Aqua-Aerobic Systems Inc.
Aquateam–Norwegian Water Technology Centre A/S
ARCADIS
Associated Engineering
Black & Veatch
Blue Water Technologies, Inc.
Boyle Engineering Corporation
Brown & Caldwell
Burgess & Niple, Ltd.
Burns & McDonnell
CABE Associates Inc.
The Cadmus Group
Camp Dresser & McKee Inc.
Carollo Engineers Inc.
Carpenter Environmental Associates Inc.
CDS Technologies Inc.
CET Engineering Services
Chemtrac Systems Inc.
CH2M HILL
CRA Infrastructure & Engineering
CONTECH Stormwater Solutions
D&B/Guarino Engineers, LLC
Damon S. Williams Associates, LLC
Earth Tech Inc.
Ecovation
EMA Inc.
Environmental Operating Solutions, Inc.
Environ/The ADVENT Group, Inc.
Fay, Spofford, & Thorndike Inc.
Freese & Nichols Inc.
ftn Associates Inc.
Gannett Fleming Inc.
Garden & Associates, Ltd.
Geosyntec Consultants
GHD
Golder Associates Ltd.
Greeley and Hansen LLC
Hazen & Sawyer, P.C.
HDR Engineering Inc.
HNTB Corporation
Hydromantis Inc.
HydroQual Inc.
Infilco Degremont Inc.
Jacobson Satchell Consultants, Inc.
Jacques Whitford NAWE, Inc.
Jason Consultants LLC Inc.
Jordan, Jones, & Goulding Inc.
KCI Technologies Inc.
Kelly & Weaver, P.C.
Kennedy/Jenks Consultants
KMK Consultants
Komline Sanderson Engineering Corporation
Larry Walker Associates
Limno-Tech Inc.
The Low Impact Development Center Inc.
Malcolm Pirnie Inc.
Material Matters
McKim & Creed
Metcalf & Eddy Inc.
Monteco Corporation
MPR Engineering Corporation, Inc.
MWH
NewFields Water Resources, LLC
O'Brien & Gere Engineers Inc.
Odor & Corrosion Technology Consultants Inc.
Original Engineering Consultants, Ltd.
Oscar Larson & Associates
Parametrix Inc.
Parsons
Post, Buckley, Schuh & Jernigan
Praxair, Inc.
RMC Water & Environment
Ross & Associates Ltd.
Rothberg, Tamburini & Windsor, Inc.
SAIC
Savin Engineers
Siemens Water Technologies
Stantec Consulting Inc.
Stearns & Wheler, LLC
Stone Environmental Inc.
Stratus Consulting Inc.
Synagro Technologies Inc.
Tetra Tech Inc.
Trojan Technologies Inc.
Trussell Technologies, Inc.
URS Corporation
Wade-Trim Inc.
Westin Engineering Inc.
Weston Solutions Inc.
Woodard & Curran
Zenon Environmental Inc./GE Water Process Technology
Zoeller Pump Company

INDUSTRY

American Electric Power
American Water
ChevronTexaco Energy Research & Technology Company
The Coca-Cola Company
Dow Chemical Company
DuPont Company
Eastman Chemical Company
Eli Lilly & Company
Merck & Company Inc.
Premier Chemicals LLC
Procter & Gamble Company
Thames Water Utilities
Severn Trent Services Inc.
Suez Environnment
United Water Services LLC

Note: List as of 7/10/07

WERF Board of Directors

WERF Research Council

IWA Publishing's authorised EU representative for General Product Safety Regulations is Diane D'Arras, 15 rue Duret, 75116 Paris, France, e-mail: safety@iwap.co.uk.

Printed and bound by CPI Group (UK) Ltd, Croydon, CR0 4YY
12/05/2026
02108876-0001